INVESTMENTS IN TECHNOLOGY

Investments in Technology

Corporate Strategies & Public Policy Alternatives

Barry Bozeman
and
Albert N. Link

PRAEGER

PRAEGER SPECIAL STUDIES • PRAEGER SCIENTIFIC

Library of Congress Cataloging in Publication Data

Bozeman, Barry.
Investments in technology.

Bibliography: p.
1. Technological innovations—United States.
2. Research, Industrial—United States. 3. Research, Industrial—Government policy—United States. I. Link Albert N. II. Title
HC110.T4B69 1983 338'.06 83-16135
ISBN 0-03-062412-6 (alk. paper)

FOR OUR PARENTS

Published in 1983 by Praeger Publishers
CBS Educational and Professional Publishing
a Division of CBS Inc.
521 Fifth Avenue, New York, NY 10175 USA

3456789 052 987654321

Printed in the United States of America
on acid-free paper

Preface

From our many discussions while attending and participating in the Conference on Technology Transfer held at the 1980 International Institute of Management Science in Berlin, Germany, it soon became clear that we shared similar interests in science, technology, and innovation. From then until now our ideas have matured into this product. The following pages reflect our understanding of the various aspects of research and development (R&D) strategies used by U.S. manufacturers for acquiring and utilizing technical information, the ways in which public policies appear to be influencing these activities, and the ways in which knowledge about these R&D activities could influence future public policy.

Many individuals and organizations have directly and indirectly contributed to this project. The financial support provided to Albert N. Link by the National Science Foundation, Division of Policy Research and Analysis, was invaluable for conducting the research summarized in Chapters 2 through 5. Barry Bozeman gratefully acknowledges the support of the Center for the Study of Science in Society and the Department of Management of Virginia Polytechnic Institute and State University where he was a visiting professor during the period in which most of this book was written. Also, our sincere thanks go to Edwin Mansfield of the University of Pennsylvania for his continual encouragement during the early stages of much of the empirical work summarized herein. In addition, Greg Tassey of the National Bureau of Standards, Eleanor Thomas of the National Science Foundation, and Nestor Terleckyj of the National Planning Association provided invaluable comments on virtually all of the papers preliminary to this work. Finally, Zvi Griliches of Harvard, John Moroney of Texas A&M, Richard Nelson of Yale, Devendra Sahal of New York University, Ken Ruscio of the U.S. General Accounting Office, Bob Zmud of the University of North Carolina, and my colleague Robert Connolly offered many useful suggestions and ideas during the various stages of the research.

We would also like to thank the editors who graciously granted us permission to include in this volume parts of previous works originally published in *The Bell Journal of Economics*, *Decision Sciences*,

Economics Letters, IEEE Transactions on Engineering Management, Managerial and Decision Economics, Southern Economic Journal, and *The Transfer and Utilization of Technical Knowledge.*

Our thanks also go to John Lambert, former editor at Praeger, for his encouragement and trust during the initial development of this project, and also to both Betsy Brown, now editor at Praeger, and Susan Yuran, Project Editor, for their skillful supervision during the final preparation of the book. Becky Askew and Vicki Sparrow, both of the Economics Department at the University of North Carolina at Greensboro, provided peerless typing and editorial assistance through the many drafts of this manuscript. To them, and to all of those mentioned above, our sincere appreciation.

Most important through the development of this book have been our wives, with their patience and encouragement, and our children, who have helped us maintain our sense of humor.

Contents

1
Investments in Technology: An Overview

INTRODUCTION

As the nation seeks to remedy its declining rate of productivity growth, a diverse assortment of prescriptions are being offered. Some of these prescriptions are "holistic medicine" and involve the creation of comprehensive industrial (or reindustrialization) policies (Wachter and Wachter 1981). Others (Harrison and Bluestone 1982) are less sanguine about the practical value and feasibility of sweeping industrial policies, especially policies including plans to prop up industries that no longer seem viable in today's changed (and changing) economy. More specific prescriptions are at least as numerous as the many alleged symptoms of the productivity decline.

While we recognize that factors as diverse as changes in workers' attitudes, increased proportions of professionally managed companies, and outmoded production equipment may have played a significant role in declining productivity, our focus is more restricted. We are concerned with technological change as it relates to industrial research and development, R&D. We do not assume that industrial R&D is a likely panacea, only that it is one of many factors that contribute to productivity and, despite the attention R&D has received from researchers and policy makers, a factor that remains poorly understood. Indeed, it is our contention that there is little reliable knowledge about industrial R&D and that public policies designed to stimulate R&D are often grounded in faith (or desperation) rather than in careful reasoning. Also, knowledge about the composition of industrial R&D, its effects on innovation and productivity, is so limited that policies are only rarely anchored in reasonable expectation. Such knowledge as has been developed concerning the dynamics of technological change has been drawn at such a general level (theories about the role of technology within the economy) or at such a specific level (case studies), that it is of limited use for policy guidance.

Economists have long recognized the economic importance of technological change. Joseph Schumpeter is frequently singled out as one economist who influenced our thinking about the role of technology in facilitating society's advancement. However, earlier recognitions can be traced as far back as Adam Smith. It was common among the earlier writers on this subject to view technology as an exogenous force: their intent was not to examine the origin of technologies, but rather to model the macroeconomic consequences of these changes. Within the past two decades much has been learned about technology—its causes and its consequences—yet there remains much that is still unknown.

Our intent in writing this book is to offer some beginnings of a microeconomic framework for thinking about the role of technology within the firm. Very little work has been done to examine systematically the sources of technology for a firm and the resulting economic consequences of these investments. This volume is one attempt to fill that void.

TECHNOLOGY AND INNOVATION AS CONCEPTS

An Informational View of Technology

Researchers concerned with technology and innovation follow a path that is well trodden. The intellectual footprints of others often help one find the way; but the traffic in technology and innovation studies can confuse. The concepts technology and innovation have been used in so many different ways that conceptual signposts are often ambiguous.[1]

The concept "technology" has been used in a relatively narrow sense to describe physical tools of a certain sort, but it has also been used to describe a wide spectrum of social processes that may be viewed as intangible tools (Sahal 1981). While there are analytical advantages to both the more restrictive and the more encompassing concepts of technology, these quite different usages almost invariably promote confusion. For our purposes, we eschew conceptualizations of technology that embrace social or other intangible entities. In our analysis such concepts as "organizational technology" (Thompson 1963) and "technological ethic" are out of bounds. But problems remain. Even if we limit ourselves to analyses of physical technologies, two conceptual problems become evident. First, how do we differentiate technologies? Second, what particular aspects of technology are of interest?

Our answer to the last question is, in a sense, an answer to the first. Our interest is in an intangible, and only indirectly perceptible, aspect of man-made physical tools. We are chiefly interested in the *knowledge* embodied in technology. Any viable technological device may be viewed as a proof (of sorts) of the informational assumptions that lead to its creation. The information embodied in a technology is diverse in regard to type, source, and instrumental application. For example, some of the information (there is much debate as to just how much) is typically derived from well corroborated scientific theory. But knowledge generated by scientists in pursuit of theory is rarely, if ever, sufficient for the more particular needs entailed in the construction of a technological device. Other sources of knowledge include information from controlled experimentations (of a focused or unfocused nature), information from trial and error coupled with incremental and sometimes minute modifications, information of the kind that philosophers sometimes refer to as "ordinary knowledge", and, finally, information of the kind that goes under the rubrics of creativity, genius, and inspiration.

It is not possible in most instances to dissect completely the information base for a technology. Often much of the knowledge reflected in a technology is deeply imbedded, so much so that it is at the nexus of scientific information and ordinary knowledge. For example, information represented in Newton's laws of motion are implicit elements of many technologies, but this does not imply that the creators of technology read Newton. A further complication arises from the fact that many technologies rely not only on deeply imbedded knowledge, but also on knowledge imbedded (and not reexamined) in related or earlier generation technologies.

Moreover, it is not our purpose to classify the information constituents of technology. Such a task is well beyond our means. Nevertheless, there are advantages to viewing technology as physical representation or proof of informational assumptions. Such a view highlights the role of research (including not only basic research, but also applied research and development) in the generation of technology. An informational view also might be used to distinguish among particular technological devices and classes of technology. One might set distinctions according to the information apparently imbedded in technologies. Admittedly, such an approach is limited. It is not, for example, easy to account for less formal and more idiosyncratic knowledge, but if one is interested in understanding the contributions of R&D to technology (as opposed to, say, understanding the morphology of technological devices), an informational view of technology appears useful. As Freeman (1977) notes, there is much confusion resulting from the fact that technology sometimes refers to a body of knowledge about techniques and, at other times, the actual physical entity. Our concern with technology as the physical representation of information provides a synthesis of these views and a synthesis that would appear useful for those concerned with the role of R&D in the creation of technologies.

Concepts of Innovation

The concepts commonly used in connection with innovation are deceptively simple. "Invention" is the creation of something new. An invention becomes an "innovation" when it is put into use. Innovations are "diffused" through the population of potential adoptors, and "adoption" occurs when a member of this population acquires the innovation. But these simple concepts are often difficult to apply.

At a most basic level, innovation has two distinct meanings. First, it is a new idea (sometimes embodied in a technological device, process or product). But sometimes the term is used to refer to a process, the act of innovating. This dual usage leads to ambiguity, albeit a kind of ambiguity that is fairly commonplace—a so-called "process-product ambiguity." A similar ambiguity is common in the use of the term "science." In using the term, one may refer to the activities the scientist engages in to produce knowledge (science as a process) *or* to the body of knowledge the scientist produces (science as a product).

Our use of the term innovation is restricted to its product meaning. However, certain terminological problems remain. An innovation is a complex entity not easily classified by a single discrete attribute. Unfortunately, there are no clear-cut criteria for distinguishing the diagnostic characteristics of an innovation. There are no generally accepted rules that permit one to say: "Yes, X is an innovation but, no, Y does not qualify." Most contributors to the diverse and fragmented literature on innovations argue that an essential ingredient of innovation is newness. According to Barnett (1953, p. 7), an innovation is "any thought, behavior or thing that is new because it is qualitatively different from existing forms." But there is considerable disagreement regarding the element of newness. At one pole are those who suggest that an innovation is something new in the world, quite dissimilar to other known entities. However, a large contingent of researchers use the term innovation to refer to an entity that is new to the adoptor (Rogers and Shoemaker 1971; Zaltman, et al. 1973).

Another important distinction is between invention and innovation. Mohr (1969) defines an invention as something new brought into *being*, whereas an innovation is something new brought into *use*. This is a helpful distinction, but one must remember that the boundaries between the two are not easily drawn. For example, most inventions are modified when brought into use. At what point is modification sufficient to produce a new invention?

For our present purpose, we view innovations as a product (rather than a process) and follow Mohr's somewhat problematic distinction between innovation and invention. Further, by our use an innovation is not simply new to the adoptor, but is a novel entity that has been brought into use. This conceptualization is useful for our purposes—understanding the role of R&D in the production of technologies.

TECHNOLOGICAL STRATEGIES OF THE FIRM: A CONCEPTUAL FRAMEWORK

Our concern is with the sources of firms' technical knowledge (how and where it is acquired) and the impact of this knowledge on firms' production processes. The starting point for our conceptual framework is to view the firm as operating in a market environment that influences, among other things, its innovative activities. Market environments differ in many respects. Markets are traditionally characterized in an economic sense on a pure competition to monopoly spectrum. This polarization is somewhat useful when focusing on price competition, but its applicability diminishes when technological competition is considered. The technological environment of a firm can be described in many dimensions. For example, in mature markets or industries where product competition is in terms of price, a firm's technology response to market pressures is likely to be toward process innovations to reduce operating costs. In contrast, in emerging markets or industries, where competition is product related, the technology response of a firm is probably toward product development innovations to improve quality or design. This linkage between market structure and production characteristics is denoted in Figure 1.1. Not only do we conceive of market competition and industry structure as directives on the firm's strategic plan, but also we acknowledge the importance of public policy constraints in the choice of technology design and source.

Figure 1.1
A Conceptual Model of Firms'
Technological Strategy and Its Determinants

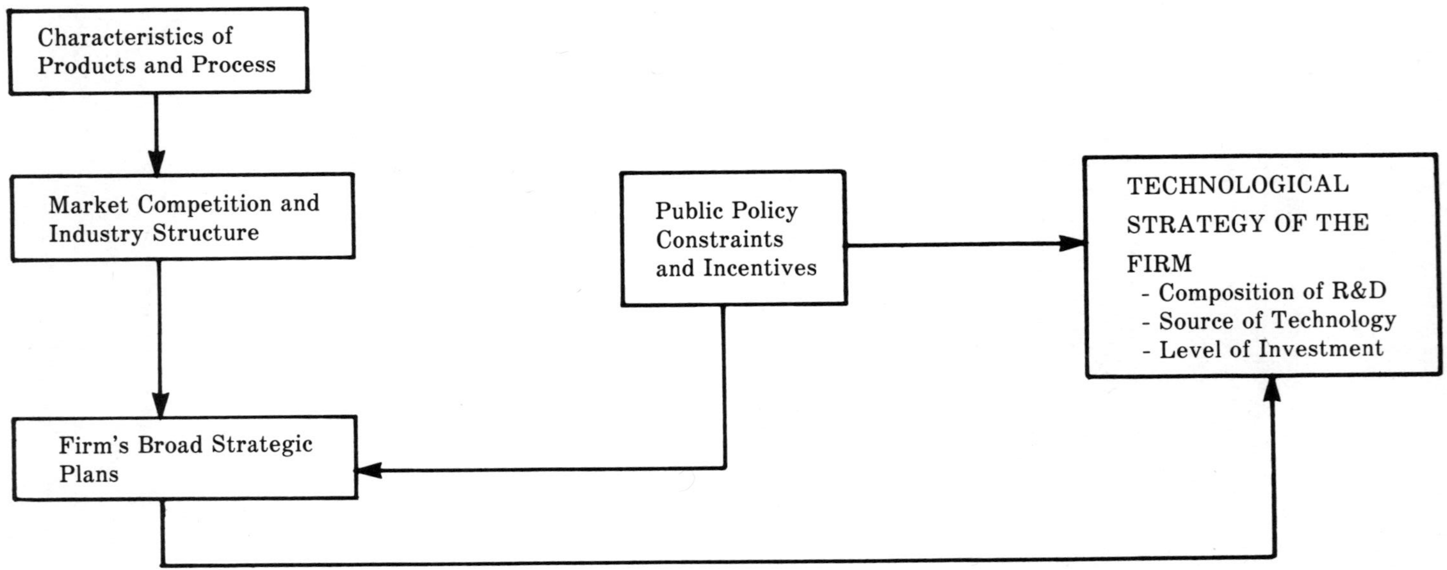

Source: Compiled by the authors.

One way the firm, or more precisely those who collectively manage the firm, responds to its environment is the way in which they acquire new technologies. A relevant strategy decision associated with this environment induced innovation stimulus is where to invest in technology. There are two choices (not mutually exclusive): the firm can internalize the innovation process by pursuing specific types of activities, broadly categorized under the rubric of R&D, or can use existing markets to purchase the requisite technology.[2]

The choice to "make or buy" can be shown to be influenced by the firm's market environment. For example, in an infant market where competition is product-related, the appropriability of technical knowledge is crucial to the successful introduction of a new design. When appropriability is less important, possibly for production process-augmenting changes, external markets can adequately provide the needed technologies.

Not only does the firm's market environment influence its overall "make versus buy" strategy for acquiring technology, it also influences the composition of its internal R&D activities. R&D encompasses myriad activities ranging from basic research to product development on one spectrum, from process innovations to product improvements on a second spectrum, or from long-term risky research to short-term applied ventures on a third spectrum. Firms differ in their R&D mix in response to their market-driven needs.

Whatever the type of R&D, or more generally, whatever sources used by the firm to acquire technology, one objective is productivity growth.

OUTLINE OF THE BOOK

As the title of this book suggests, our focus is on corporate investments in technology. In general we examine firms' strategies for acquiring technical knowledge—the determinants and the consequences—and we investigate current public policy responses to alter both the magnitude and the direction of the investment patterns. (Again, refer to our model in Figure 1.1.) Throughout, we make use of survey data from U.S. manufacturing firms collected by Link over the past five years.

In Chapter 2, "Alternative Sources of Technology," we develop the idea that there are several alternatives available to a firm for acquiring technical information—the firm's own R&D program and external technical markets. We speculate as to the nature of each and as to the incentives facing firms to allocate their resources between each. Empirical evidence is presented to document the disparate corporate strategies for acquiring technologies among U.S. manufacturing firms, and hypotheses are tested to explain inter-firm differences in this behavior.

In Chapter 3, "Corporate Investments in R&D," we document the changing trend in R&D spending during the past decade away from the basic, long-term risky ends of the R&D spectrum toward the development and short-term applied ends.

In Chapter 4, "An Analysis of the Composition of R&D Spending," we offer several hypotheses regarding the characteristics of firms that influence the allocation of R&D resources among the various categories of spending. The empirical analysis is related to inter-firm differences in process versus product innovation spending and on basic research versus applied research versus development spending.

In Chapter 5, "Productivity Growth and Investments in Technology," we examine the relationship between productivity and technology. More specifically, we consider the sources and kinds of technology used by firms in production and evaluate their impact on measured productivity growth. A detailed analysis of the components of R&D and their relation to productivity is also provided.

In Chapter 6, "Tax Incentives for R&D: A Critical Evaluation," existing and proposed tax incentive policies are examined. There are three fundamental questions addressed. What is the need for R&D tax incentives? What is the appropriate policy tool (grants assistance, subsidy, or tax incentives)? What are the design criteria by which one might evaluate a tax incentive? In addition to examining the apparent economic efficiency of various approaches, we also give some attention to political and administrative questions.

A particular tax incentive proposal is examined in Chapter 7, "The Economics of a Tax Credit for Joint Venture R&D." We argue that a joint venture tax credit would in all likelihood generate an increased proportion of basic research, an outcome that we, like others, view as socially desirable. A formal theoretical framework is presented that has implications for determining the likely effects of a joint venture tax credit.

Finally in Chapter 8, "Industrial R&D and Public Policy," we summarize our findings and suggest several possibilities for directions of future research in this area. Our focus is on research topics that might be especially relevant for informing, and hence focusing, public policy.

NOTES

1. The following discussion of the conceptional problems associated with the use of the terms "technology" and "innovations" draws, in part, from Loveless and Bozeman (1983).
2. More precisely, firms develop technological strategies that include some mix of induced and purchased technologies.

2

Alternative Sources of Technology

INTRODUCTION

As discussed in Chapter 1, we conceptualize technology as the application of knowledge. With individuals, knowledge comes from myriad sources. Formal education is one form of acquiring knowledge, but knowledge can also be gained experientially. Each of these sources may augment the human capital of individuals differently. So it is with a firm's technical knowledge. There are many alternative sources open to a firm for acquiring technical information.

Broadly speaking, one can dichotomize these sources as either internal or external to the firm. The most obvious internal source of technology is the firm's indigenous R&D activity. Although much has been written about investments in R&D – their determinants and their consequences – one should not assume that indigenous R&D activity is the most important source of technology for all firms. External sources of technical knowledge are also important. One source is the federal government. Firms involved in contracted research can appropriate knowledge from their project involvement. In addition, private-federal interactions enrich the technical environment of the firm by establishing new networks for the future exchange of ideas.[1] A second external source of knowledge is other firms, both within the industry and outside of the industry. New technologies are embodied in equipment and can enter a firm through the purchase of newer

---

An earlier version of the material presented in the last two sections of this chapter first appeared in *Decision Sciences* (Link, Tassey, and Zmud 1983).

vintages of capital produced by other firms. Also technologies can be licensed from other firms, or acquired from other firms through mergers.

In the following section of this chapter we document, using survey data, differences in the relative importance of each of these alternative sources of technology. Then, we suggest a theoretical explanation of why firms differ in their strategies for acquiring technology. Finally, the implications of our model are tested empirically.

THE IMPORTANCE OF ALTERNATIVE TECHNOLOGY SOURCES

We present here the results from two surveys of R&D active manufacturing firms. In the first survey, conducted in 1978 (hereafter referred to as the 1978 survey), information was obtained from 329 firms regarding the percentage of their technological advances that resulted from indigenous (self-financed) R&D activities. We will call these kinds of advances *induced innovations.* (The 329 sample firms accounted for 71.2 percent of total 1977 net sales in the manufacturing sector and 74.1 percent of the total 1977 company-financed R&D conducted within the manufacturing sector.) Presumably, if innovations did not arise as the result of R&D-related effort, then the innovations must have been developed outside the firm and must have entered the firm through such channels as the licensing of others' technology or the purchasing of capital equipment that embodies new technology (perhaps the result of a capital supplier's R&D). We will call these kinds of advances *purchased innovations.*

From our data there appears to be a significant degree of inter-firm and inter-industry variability in the percentages of technological advances that are induced. To illustrate this variability, the mean percentages (averaged across firms) of induced technologies for 11 separate industry groupings are presented in Table 2.1. These percentages range from a low of 25 percent in the textile and apparel industry to a high of 68 percent in the chemicals industry. For the entire sample of 329 firms the range of reported values is 12 percent to 83 percent, with a mean of 48 percent and a standard deviation of 31 percent.

One implication from the data is that indigenous R&D spending is not the only source of technical knowledge for firms in the manufacturing sector. This conclusion does not necessarily suggest that the importance of R&D activities as a source of innovation varies in proportion with the percentages shown in Table 2.1. For example, firms in the textiles and apparel industries appear to purchase, rather

TABLE 2.1. Selected Industry Means of Percentage of New Technological Advances Induced Through R&D

Industry	*Mean Percentage*
Food and kindred products (SIC 20)	46
Textiles and apparel (SIC 22, 23)	25
Paper and allied products (SIC 26)	57
Chemicals (SIC 28)	68
Petroleum refining (SIC 29)	49
Primary metals (SIC 33)	36
Fabricated metal products (SIC 34)	56
Machinery (SIC 35)	34
Electrical equipment (SIC 36)	43
Transportation equipment (SIC 37)	43
Professional and scientific instruments (SIC 38)	38
Manufacturing Sector	48

Note: SIC – Standard Industrial Classification.

Source: 1978 survey of 392 firms by authors.

than induce, the larger portion of their innovations. This finding does not imply necessarily that the technological importance of those innovations purchased is greater than those innovations induced from indigenous R&D spending.

Having documented that alternative sources of technical information are widely used by manufacturing firms, an attempt was made to determine the relative importance of certain technology sources. In a 1982 follow-up survey of these 329 manufacturing firms (hereafter referred to as the 1982 survey) data were collected from 146 firms regarding the relative importance of six alternative sources of technology. This subset of 146 firms accounted for 31.2 percent of total 1979 company-financed R&D in the manufacturing sector. Approximately 83 percent of these firms spent more than $10 million on R&D during 1979.

Each of these 146 firms was asked to rank, on a five-point Likert scale ranging from "very important" (= 1) for enhancing innovative activity to "not important at all" (= 5) for enhancing innovative activity, each of the following six sources for technical information: indigenous R&D activity, R&D conducted under government contract, new capital equipment purchases, technologies licensed from other domestic firms, technologies licensed from other international firms, and technologies acquired from other firms through mergers. The mean responses from the sample of firms are presented in Table 2.2.

TABLE 2.2. Mean Ranking of Alternative Sources of Technology

Source	*Mean Ranking*
Indigenous R&D activity	1.3
R&D conducted under government contract	4.2
Purchasing new capital equipment	2.7
Licensing from domestic firms	3.7
Licensing from international firms	3.8
Acquisition of firms through mergers	3.1

Source: 1982 survey of 146 firms by authors.

Although firms make extensive use of non-R&D related innovations, they appear to consistently view their own R&D efforts as being relatively more important than any of the other sources for enhancing their innovative activities. As can be seen from Table 2.2, the mean ranking of indigenous R&D is 1.3 on a scale from 1 to 5. Second in the ranking is the purchasing of new capital equipment. Least important is government-contracted R&D. The mean response of 4.2 is from only those firms that receive, or have recently received, federal grants or contracts.

These responses were also examined by industry. Using both two-digit and three-digit Standard Industrial Classification (SIC) groupings, indigenous R&D activities consistently ranked as the most important source. However, variations in the relative ranking of the other sources occurred across industries. In the computing equipment industry (SIC 357) and in the professional and scientific instruments industry (SIC 38), mergers and licensing from domestic firms were both reported to be the next most important sources for complementing firms' innovative activities. These mean responses are reported in Table 2.3.

TABLE 2.3. Importance of Mergers and Domestic Licensing in Computing Equipment and Scientific Instruments Industries

	Computing Equipment (SIC 357): n = 21	*Scientific Instruments (SIC 38): n = 15*
R&D		
Mean Ranking	1.4	1.1
Correlation with size	-.11	-.17
Mergers		
Mean Ranking	2.1	2.6
Correlation with size	-.36	-.27
Domestic Licensing		
Mean Ranking	2.7	3.0
Correlation with size	+ .29	+ .41

Note: SIC – Standard Industrial Classification.

Source: 1982 survey of 146 firms by authors.

In both industry groups, mergers ranked as the second most important source for acquiring new technology. We also examined within each of the two industry groups the relationship between firm size (measured as 1981 net sales) and the ranked importance of each of the three technology sources. The simple correlation coefficient between size and the ranked importance of indigenous R&D is negative. This negative correlation is interpreted to mean, given that a ranking of 1 refers to "very important" and 5 refers to "not important at all," that larger firms in these two industries view indigenous R&D as relatively more important as a source of technology than do smaller firms. However, the correlation coefficients are not statistically different from zero at conventional levels. The more interesting finding is in regard to mergers and domestic licensing. From our analysis it appears that in both the computing equipment and scientific instruments industries, larger firms have relied more on merger activities (presumably merging with smaller firms) and smaller firms rely more on domestic licensing (presumably licensing from larger firms) as sources for acquiring technical information.

THE DECISION TO INDUCE OR PURCHASE TECHNOLOGY

It is clear from the information presented in the previous section that firms rely heavily on alternative sources for acquiring technical information. The most comprehensive, and perhaps the most accurate, data are those from the 1978 survey that are summarized in Table 2.1. Those data suggest that there are inter-firm differences in the percentage of new technology induced through indigenous R&D activities.

In a broad sense, one may view these firms' decisions to invest in their own R&D program — that is, to induce or make technology — as a conscious plan to internalize an activity for which there already exists an alternative market mechanism. This issue of internalizing a market activity was first addressed by Coase (1937), and later extended theoretically by Alchian and Demsetz (1972), Teece (1980, 1981), and Williamson (1975, 1979), and empirically by Mowery (1981). One conclusion from these works is that inter-firm differences in the percentage of new technology induced through a firm's own R&D efforts reflect inter-firm differences in the perceived net benefits from internalizing the innovation search. In other words, a firm will choose, at the margin, the make or buy alternative that is cost efficient in net terms.

There are benefits and costs to each alternative in the make versus buy decision. Consider first the firm's decision to purchase a given technology product from a supplier. One obvious cost is the purchase price; however, other costs arise if the supplier's product fails to meet the firm's specific requirement set. These added costs reflect, among other things, expenses involved in modifying the purchased product to meet a specialized requirement or the opportunity costs associated with relaxing a particular requirement to meet the specialized designs of the purchased product. A rational supplier will market a technology only when a demand for that technology exists. In order to enlarge potential demand, the technology product will be designed to meet the generalized needs of a number of manufacturing firms. While some flexibility (regarding the attributes of the products to be manufactured, the linkages with in-place manufacturing processes, and the precise functions to be performed) will be incorporated in the design, in many instances the successful adoption of the technology by a firm occurs only after considerable adaptation. Also, purchasing technology is always accompanied by the possibility that a firm's internal capability to innovate effectively with regard to its manufacturing processes will be diminished, thus weakening the firm's long-run technologically based competitive stance within its industry.

There are, however, benefits to purchasing technology. Purchased technologies often reduce technological uncertainty within a firm. This uncertainty refers, in part, to a lack of understanding regarding the most appropriate production process designs for handling a given manufacturing activity. Much of this uncertainty reflects the fact that a design is essentially an abstraction at its inception. Only by observing, examining, or experimenting with an evolving or an existent design can technological uncertainty be reduced (Ettlie and Vallenga 1979).

Regarding the costs of inducing technology, the direct costs of inducement replace the price component of the purchased alternative. These inducement costs generally would be expected to exceed the sale price of a purchased technology alternative since the supplier of that alternative could spread development expenses among all purchasers. These direct inducement costs may be compounded by the technological uncertainties that the firm must incur as it begins the inducement process with an abstract idea and attempts to develop the idea into a workable process design (Utterback and Abernathy 1975).

However, just as there are benefits to purchasing a technology there are benefits to inducement. First, the output from inducement is likely to better correspond to the firm's technical requirement set. Second, the firm can often appropriate the technical knowledge that arises during the innovation process. This ability to appropriate technical knowledge increases in importance to the extent that R&D is a form of nonprice competition. Nelson (1959) proffered a related idea more than two decades ago regarding one aspect of R&D, namely basic research activity. He argued then that the greater a firm's capacity for utilizing technical knowledge, the more likely it is to invest in basic research.

Firms may differ in the way they perceive benefits and costs. For example, the costs associated with internally-developed technology may be weighted less heavily if the firm in question already possesses the requisite financial capabilities and technical facilities for successful R&D. Also, if the firm's climate is supportive of risk taking, or if it places a high relative value on innovation, then the potential uncertainties associated with inducement relative to the expected benefits from an investment in a purchased technology should diminish.

Firms' perceptions about the benefits and costs, or about opportunities (technical or otherwise) may vary depending on, among other things, the entrepreneurial talents within the firm. While there is no single, universally accepted concept of the entrepreneur, history suggests the successful entrepreneur possesses unusually keen perception (Hébert and Link 1982). Although the seeds for this particular entrepreneurial talent can be found as far back as in the writings of Clark (1892), more recently Schultz (1975) has suggested that entrepreneurship is a significant element in a human capital approach to understanding both market and nonmarket activities. Specifically, Schultz sees the entrepreneur as one who perceives and reacts to disequilibria. Clearly, perceiving technological opportunities in external markets is a necessary, but not sufficient, condition for successful participation in these markets.

The approach taken here in formulating a model of the induce versus purchase, or make versus buy, decision begins with the assumption that the benefits and costs from inducing or purchasing are largely a function of the firm's manufacturing environment. Accordingly, the environment presents firms with differing incentives for acquiring technologies in various ways. One representation of the manufacturing environment that seems relevant to innovation, process innovation in particular, is the Utterback and Abernathy (1975) product-process life cycle model. Although some (for example, Pavitt and Rothwell 1976) have criticized this formulation, it remains intuitively appealing and widely used (Ettlie 1979; Hayes and Wheelwright 1979; Rogers 1982).

Utterback and Abernathy suggest that the induce versus purchase decision is related to the firm's production process life cycle:

> Characteristics of the [firm's internal] innovation process . . . will vary systematically with the firm's environment . . . *and* [emphasis added] with the state of development of process technology used by a firm and its competitors. (1975, p. 640)

Major product innovations often arise in small firms characterized by a unique technical capability, market understanding, or both. As a product attains market success, production volume increases thus stimulating the need for new process technologies, other firms enter the market, and cost competition via process technologies commences. Because these requirements for new process technologies arise as a result of evolutionary forces, internal development of new process technology peaks during this transition stage in the life cycle. As a product line fully matures in the specific stage, product design stability and intensive price competition lead to the use of fully integrated and automated production processes. In such endeavors economies of scale increase in importance and the firms competing successfully in the market tend to be large. With the emergence of a dominant process design, a supplier industry develops. The significant expenses and risks associated with changing an integrated, automated production process tend to weight heavily the technological uncertainty of inducing new process technology in the specific stage, and the relative advantage of purchasing this technology increases. These ideas are summarized in Table 2.4.

The original Utterback-Abernathy framework is a descriptive model of production process development within a single plant that anticipates a common pattern of change among different industries. This expectation of uniformity across firms has been criticized along with the model's limited focus; namely, a single plant. Ettlie (1979) has recently proposed that the Utterback-Abernathy model may be most appropriate when a firm and its environment move together in development.

Recent conceptual advances in organizational theory adopt a similar position to that of Ettlie (1979). Brittain and Freeman (1980), for example, suggest that while a given industry may simultaneously exhibit a number of production process life cycles, only one – that most closely linked with the industry's generic underlying technology – will dominate. They write:

> The basic ecological dynamic prevails in all organizational populations, with the outcome in any industry specific to the environmental conditions under which organizational forms seek to exercise their relative advantages. (1980, p. 338)

Very similar notions can also be found in the views of Hage (1980). After examining the question of whether organizations choose strategies or whether environments constrain strategic choices, Hage concludes that over the long term most strategies are institutionally rather than organizationally determined:

> The argument running through this whole discussion is that in general there is not much strategic choice. *Given certain constraints that determine the resources and performances of the organization, then there are appropriate . . . strategies.* (1980, p. 451)

The constraints alluded to by Hage involve the technological, economic, and political forces that operate within an organization's specific

TABLE 2.4.
Key Elements of the Production Process Life Cycle Model

Stage of Process Development	*Product Characteristics*	*Process Characteristics*	*The Induce Versus Purchase Decision*
Fluid	• rapid, frequent product changes • product diversity • high margins	• rapid, frequent process changes • process inefficiency • labor-intensive • general purpose process equipment • use of available materials	• reluctance to invest R&D funds into process innovation • purchase of general purpose equipment and available materials
Transition	• maturing product group • emergence of dominant product design	• islands of process specialization arise • process development becomes important to increase output volume and to meet cost competition	• extensive internal process development • lack of a supplier industry
Specific	• product standardization • low margins	• emergence of dominant process design • integrated automated process technology • capital-intensive • process changes major and costly	• emergence of a supplier industry • purchase of specialized process technologies

Source: Albert N. Link, Gregory Tassey; and Robert W. Zmud. 1983. "The Induce versus Purchase Decision: An Empirical Analysis of Industrial R&D." *Decision Sciences* 14:46-61.

environment while the organizational performances refer to a firm's choices between quality and quantity and between innovation and efficiency.

Here we extrapolate the Utterback-Abernathy model from a plant to an industry life cycle in order to explain a firm's decision to make or to buy a new technology. Consequently, we hypothesize that the propensity of firms within an industry to induce technology will roughly follow an inverted-U shape: firms in industries categorized as being at the fluid or specific stage of process development will tend to purchase relatively more new technology, whereas firms in industries at a transitional stage will tend to induce their technology. Therefore, industry characteristics likely to influence a firm's decision to induce or purchase new process technology would be those descriptive of an industry's stage of process development.

In the following section we test the hypothesis that the incentive for firms to induce technologies via their own R&D activities is related to the industry's stage of process development using the data summarized in Table 2.1.

THE ANALYTICAL FRAMEWORK

The theoretical discussion in the previous section suggests that inter-firm differences in the decision to induce or purchase new technologies may be systematically related to certain industry characteristics that describe the dominant stage of process development. Specifically, we predict that cross-industry variations in the average firm propensity to induce technology (INDUCE) will follow an inverted-U shape along the spectrum of stages of process development from fluid to specific (see Table 2.4).

Two industry variables are considered in our model as proxies for the stages of process development. The first is the capital intensity of the industry in which a firm operates, as measured by the industry's capital to labor ratio (K/L). Industries at a fluid stage of process development tend to be labor intensive and industries at a specific stage tend to be capital intensive. Since incentives to induce new process technology are believed greatest at the transition stage of process development, we expected that industries characterized by an intermediate degree of capital intensity would induce rather than purchase technology, *ceteris paribus*.

The second variable is an index of industry product standardization (STAND). This variable is employed to capture the extent to which an industry's firms are able to exercise competitive autonomy.

With high product standardization, innovation targeted at product innovation realizes little gain as the risk associated with introducing new products is high. Furthermore, the emergence of a dominant product design is likely to be accompanied by the evolution of a supplier industry marketing manufacturing processes for producing the standardized product (Link 1982e, 1983b). Thus, it might be expected that firms in such an industry would purchase rather than induce new process technologies, *ceteris paribus*.

Other variables which may affect the decision are held constant to test for the influence of a firm's industry stage of process development on its incentive to induce. For example, the market structure in which a firm produces is also likely to affect its induce versus purchase strategy. Innovation is a form of nonprice competition, and nonprice competition occurs to a greater extent in markets where there is at least some degree of seller concentration. A firm in a concentrated industry (CONC) will have the incentive not only to innovate but also to appropriate as much of the technical knowledge associated with that innovation as possible. Therefore, one might expect that the incentive to induce is greater in industries rich in nonprice competition, that is, in industries with some degree of seller concentration, *ceteris paribus*.

Even after controlling for industry factors influencing the induce versus purchase decision, intra-industry inter-firm variation in the firm percentages summarized in Table 2.1 may still remain. Therefore, three firm characteristics are also included in our empirical model.

First, larger firms are expected to induce a greater percentage of their new technological improvements than smaller firms. A frequent interpretation of Schumpeter (1947) is that larger firms can more easily provide the physical and financial economies of scale in production and innovation requisite for success in the R&D process. Therefore, some of the technical uncertainties surrounding technological advancements are reduced as size increases. As we pointed out in the previous section, there are technical uncertainties associated with purchasing technology regarding the adaptability of a particular technology to the firm's production process or to its requirement set. Assuming these uncertainties remain for all purchasers, then as firm size (SIZE) increases the relative benefits of indigenous R&D also increase, *ceteris paribus*.

Second, a firm's ability to perform successful R&D (SUCC) may reduce the technical uncertainties associated with inducement. A history of successful R&D behavior may also provide impetus for innovation leadership within an industry. If so, then the option of purchasing technology is relatively less attractive, *ceteris paribus*, since the ability to appropriate the associated technical knowledge is reduced.

Third, more diversified firms (DIV) may induce to a greater degree, *ceteris paribus,* because they are better able to utilize the unexpected spillovers of technical knowledge resulting from the R&D process. This firm characteristic also reduces the costs of adaptation associated with induced activities.[2]

These hypotheses are tested using the following linear regression specification:

$$\text{(2.1) INDUCE} = \alpha_0 + \underset{(+)}{\alpha_1} \text{ K/L} + \underset{(-)}{\alpha_2} (\text{K/L})^2 + \underset{(-)}{\alpha_3} \text{ STAND} + \underset{(+)}{\alpha_4} \text{ CONC} + \underset{(+)}{\alpha_5} \text{ SIZE} + \underset{(+)}{\alpha_6} \text{ SUCC} + \underset{(+)}{\alpha_7} \text{ DIV} + \varepsilon,$$

where ε is an error term assumed to obey all classical assumptions. Below each regression coefficient is our predicted direction of influence of each variable on INDUCE.

The Data

The dependent variable, INDUCE, represents the percentage of a firm's technological advances that come from self-financed indigenous R&D activities. As discussed in reference to the summary statistics presented in Table 2.1, these data come from a 1978 survey of 329 manufacturing firms. Only 275 of these firms will be used in the estimation of equation (2.1) because of missing data on several of the independent variables.

The industry capital-to-labor ratio variable associated with each firm is measured as the ratio of gross book value of depreciable assets in millions of 1976 dollars divided by the total number of 1976 employees, in thousands. These data, at the four-digit SIC level, come from the U.S. Bureau of the Census *Annual Survey of Manufacturers.* Each firm's four-digit classification was obtained from either Compustat, or the Standard & Poors' *Register of Corporations.*

There are no published statistics related to any aspect of the degree of product standardization within an industry. A four-digit industry index of the intensity of voluntary product standards was calculated using information from the U.S. National Bureau of Standard's *Tabulation of Voluntary Standards* and *Certification Programs for Consumer Products* and from the unpublished documentation to that publication. Specifically, our index measures the proportion of seven-digit product lines within a four-digit category for which at least one voluntary standard existed by 1976 (Link 1982e). However, since this variable applies only to consumer goods industries it is omitted initially from our estimation of equation (2.1). The standardization variable is included in a second estimation using a smaller sample consisting only of firms in consumer goods industries.

Seller concentration is measured as the four-firm concentration ratio corresponding to the four-digit SIC industry in which Standard & Poors classifies each firm. The data come from the U.S. Bureau of the Census' 1972 *Census of Manufacturers.* Standard & Poors classifies some firms only by a three-digit or two-digit SIC industry code on the Compustat files. In those instances the *Register of Corporations* was used to determine a unique four-digit code.

Firm size is measured, in millions of 1976 dollars, as gross sales and other operating revenues less discounts, returns, and allowances as reported by Compustat. Each firm's success rate in previous R&D endeavors was obtained as part of the 1978 survey. SUCC represents each firm's R&D vice-president's subjective assessment of the percentage of its R&D projects that profitably reached commercialization. There are no other sources with which to measure the accuracy or representativeness of these survey responses. We attempted, through interviews, to establish inter-firm consistency in the responses; therefore, although the magnitude of SUCC may be biased, cross-firm variability in its values is presumed to have meaning.

The extent of each firm's product differentiation is represented by the number of four-digit SIC industries in which the firm and its subsidiaries operate. These data also come from the *Register of Corporations.* No attempt was made to weight this index by the percentage of the firm's sales in each of the four-digit industries. Since, as hypothesized above, one incentive to induce is provided by the firm's potential to use whatever unexpected technical knowledge results, it is the absolute number of markets served by each firm that is important in measuring inter-firm differences in potential sources of application.[3]

The Statistical Results

The least-squares regression results corresponding to equation (2.1) are presented in Table 2.5. The results reported in column (1) are for the subsample of 275 firms. As discussed above, values on the standardization variable, STAND, are available only for the subsample of 115 firms in consumer goods industries (as defined by Mueller and Rogers 1980). Those results are reported in columns (2) and (3).

The industry characteristics considered in the theoretical model performed as hypothesized. The signs on the K/L terms suggest the

TABLE 2.5.
Estimated Regression Results from Versions of Equation (2.1)
(t-statistics in parentheses)

Independent Variables	*(1)* *n = 275*	*(2)* *n = 115*	*(3)* *n = 115*
Intercept	.21	.16	.11
	(4.31)**	(2.79)**	(3.01)**
K/L	1.12	.53	.61
	(.78)	(.28)	(.31)
$(K/L)^2$	-.047	-.036	-.039
	(-1.99)*	(-2.13)*	(-2.19)*
STAND	-	-	-.491
			(-2.51)*
CONC	.037	.029	.027
	(2.77)**	(2.16)*	(2.20)*
SIZE	.00024	.00015	.00011
	(3.68)**	(3.47)**	(3.51)**
SUCC	.071	.110	.107
	(1.99)*	(2.12)*	(2.01)*
DIV	.005	.003	.003
	(3.02)**	(2.75)**	(2.81)**
R^2	.559	.605	.654

** significant at the 0.01 level
* significant at the 0.05 level
Source: Compiled by the authors.

predicted inverted-U shaped relationship, although only the estimated coefficient on the $(K/L)^2$ term is statistically significant. We interpret this finding to mean that few firms in our sample are in industries producing in the fluid stage of development. As well, it is likely that those fluid industries that are represented are not fully captured by a single four-digit K/L measure. Industry concentration is a significant factor in our empirical model. In other versions of equation (2.1) we considered a nonlinear CONC term to test for the possibility of an inverted-U shaped relationship. Such a relationship was previously found by others when analyzing different aspects of nonprice competition such as advertising and R&D spending (Greer 1971; Martin 1979; Scherer 1967a; Scott 1978, 1981; Strickland and Weiss 1976). We found a negative coefficient on the nonlinear CONC terms, but it was not significant.

Regarding the firm characteristics, the estimated coefficients on each of the variables are the predicted sign and are significant. In other estimating versions of equation (2.1) we included nonlinear SIZE and DIV terms to test for the possibility of diminishing returns on the firm's incentive to induce. There was no evidence of diminishing returns to firm size, but the positive statistical influence of DIV diminished after firms became diversified in more than 18 four-digit SIC industries (the range of DIV is 1 to 39).

The regression results reported in columns (2) and (3) of Table 2.5 correspond to versions of equation (2.1) with and without the variable STAND. The fact that firms within this subsample behave in a manner similar to that of the complete sample is evidenced by the consistency between the results in columns (1) and (2). Of particular interest is the increased size of the regression coefficient on SUCC reported in column (2). This increase probably reflects the fact that R&D activity in consumer goods firms is more development intensive (that is, less basic research) than in the rest of the manufacturing sector.

Standardization also appears to be an important determinant of induced technological activities. Its estimated coefficient is the predicted sign and is significantly different from zero. Its inclusion, as reported in column (3), accounts for nearly 5 percent of the explained variation in INDUCE.

These reported results represent the first empirical attempt to quantify the make versus buy decision across a large number of firms in the manufacturing sector;[4] however, they should be interpreted cautiously. Our dependent variable is subject to, among other things, intermanager differences in interpretation of our (or for that matter, any) survey instrument. The other variables have measurement problems too. Nevertheless, the estimated results do suggest that the theoretical constraints of the model are being at least partially measured.

CONCLUSIONS

We have argued in this chapter that firms' sources of technology are broader than their own R&D efforts. Technologies can, and are, purchased through market alternatives. This conclusion may appear to be obvious, but surprisingly very little effort has been made to determine the extent to which innovative activities occur within firms, but whose origin is from outside of firms. It is well known that innovations diffuse inter- and intra-industry but little has been done systematically to evaluate the determinants of this kind of behavior.

The results presented here support the notion that identifying alternate sources (primary and secondary) for stimulating process innovation may prove to be a viable strategy in targeting federal policies for achieving technological advance.[5] More importantly, it is apparent that with certain industries (those dominated by products in the specific stage of process development and those exhibiting high levels of product standardization) the most efficient means of encouraging process innovation may be by stimulating the R&D efforts of the industries' suppliers. Such an outcome adds emphasis to the need for research aimed at better understanding the role of suppliers in industrial R&D.

This first implication suggests a strategic perspective for firms' marketing process technologies to other firms. Decisions to offer new products or to market a major product might be enlightened by recognition of the competitive environment of targeted industries. Two markets might be involved: one involving general purpose process equipment directed toward firms competing on a basis of product quality and innovation and the other involving firms competing on a basis of product quantity and efficiency. We further advocate research investigating the nature of these markets for new process technologies since such knowledge would result in a better understanding of the strategic decision to purchase or induce new process technology.

Finally, our findings also support the recently advanced views that federal policies directed toward encouraging industrial R&D must be cognizant of the potential for differing effects across industries. We will elaborate on this policy implication in Chapter 6. Clearly, more research is needed aimed at identifying industry groupings most amenable to specific federal R&D policies.

In the next chapter we examine in detail the microeconomic nature of R&D activity. Although we have viewed R&D as one source of technology for a firm, the influence of R&D is even more encompassing since the technologies purchased are presumably the output from other firms' R&D. And R&D is not homogeneous in nature: a variety of innovation-related activities fall under the broad rubric of "R&D." These heterogeneous activities are described in Chapter 3.

NOTES

1. The relationship between federal R&D and the firm's private R&D activities will be considered in more detail in Chapter 4.
2. In a sense, as we noted earlier, this hypothesis is an extension of Nelson's (1959) proposition that diversification is a prerequisite for basic research. We explicitly test Nelson's proposition in Chapter 4.
3. Grabowski (1968) and McEachern and Romeo (1978) have used the number of five-digit SIC product lines in which a firm produces as a measure of diversification. Those data, however, were last published in Fortune's 1966 *Plant and Product Directory*. The measure of DIV used here will be used again in the analysis presented in Chapter 4.
4. Buer (1980) has examined empirically the make versus buy distinction for a small sample of firms in five select industries. While his pilot study is not formulated in the form of an econometric model as is ours, the dependent variable is captured in a similar manner. Our survey results show a striking consistency within four of his five industries.
5. This idea will be more fully developed in Chapter 6.

3

Corporate Investments in R&D

INTRODUCTION

As we emphasized in the previous chapter, firms generally view their own R&D programs as the most important source for enhancing their innovative activities. But purchased innovations presumably are also the result of others' R&D. Whenever they take place, the R&D activities form the basis for technological advancements. Therefore, a more detailed discussion of R&D activities seems warranted.

Research and development spans myriad activities. Research may be defined as the primary search for technical or scientific advancement; and development may be defined as the translation of such advancements into product or process innovations. In the 1950s the National Science Foundation fostered a more detailed breakdown: basic research, applied research, and development. Basic research represents original investigation for the advancement of scientific knowledge that does not have a specific commercial objective. Applied research represents investigation directed to finding new scientific knowledge that has a specific commercial objective – product related or process related. Development is that technical activity concerned with nonroutine problems encountered in translating research findings and general scientific information into products or processes (National Science Foundation 1982b). In practice, industrial research and development activity is more heterogeneous than these three labels imply.

Other categories besides basic/applied/development are meaningful for disaggregating R&D. For example, R&D can be divided into expenditures resulting either in the innovation of new or improved products — product innovations — or in the innovation of new or improved production processes — process innovations. Also, R&D can be described along continua ranging from risky to nonrisky, or from long term to short term.

As we demonstrate in later chapters, insight into the economic consequences of R&D can be obtained when R&D is quantified by its disaggregated character of use. However, in most empirical studies researchers continue to group these varied activities under the composite rubric of R&D. In this chapter we present some descriptive statistics on the composition of R&D.

PATTERNS OF R&D SPENDING

National Trends

There are two major sources of R&D funding in the United States. Currently, the federal government accounts for 47 percent of all national R&D spending while private industry accounts for 50 percent. The remaining 3 percent comes from universities and colleges, and from other nonprofit institutions (National Science Foundation 1982a). Industry, the manufacturing sector in particular, performs 72 percent of this R&D.

R&D spending, in constant dollars, rose steadily from World War II to about 1968. Since then, the pattern of growth has been somewhat erratic. The data in Table 3.1 show the average annual rate of change in national R&D for five selected time periods. A most dramatic decline in the rate of growth in total R&D came between 1967 and 1975. Since 1975, both industry and federally-funded R&D have grown at a rather constant rate, industrially-funded R&D has grown at a faster rate than has federally-financed R&D. The average annual rate of growth in federally-financed R&D went from 10.2 percent (1953-67) to 1.6 percent (1981-82). Likewise, R&D spending as a percent of Gross National Product has fallen and then began a recovery (not shown): this ratio peaked at 2.96 percent in 1964, then fell to 2.34 percent in 1977-78, by 1982 it has reached 2.45.

There has been a trend away from basic research toward development activities. As shown in Table 3.2, real spending on basic research has grown over time at a slower rate than has spending on either applied research or development. Basic research outlays grew faster than either applied research or development during the 1953 to 1967 time period. However, beginning in 1975 the financial commitment to R&D has been relatively more toward the development end.

One can infer from the statistics in Table 3.2 that the absolute percentage of total R&D allocated to basic research, as opposed to applied

TABLE 3.1
Average Annual Rates of Change in Constant Dollar R&D for the U.S. Economy

Years	*Total R&D*	*Federal R&D*	*Non-Federal R&D*
1953-67	9.1%	10.2%	7.5%
1967-75	-0.5	-2.7	2.7
1975-80	4.5	2.8	6.2
1980-81	2.7	1.8	3.5
1981-82	2.6	1.6	3.4

Source: National Science Foundation. *National Patterns of Science and Technology Resources, 1982* (U.S. Government Printing Office), p. 6.

TABLE 3.2
Average Annual Rates of Change in Constant Dollar R&D for the U.S. Economy by Character of Use

Years	*Basic*	*Applied*	*Development*
1953-67	12.5%	7.6%	9.0%
1967-75	-0.4	0.5	-0.8
1975-80	4.1	4.5	4.5
1980-81	-0.3	-0.1	4.2
1981-82	-2.0	0.7	4.0

Source: National Science Foundation. *National Patterns of Science and Technology Resources, 1982* (U.S. Government Printing Office), p. 3.

research or development, has fallen over the past 30 years. As seen in Table 3.3, the percentage of total R&D going to basic research peaked in 1971 at 13.77 percent of total R&D, and has yet to grow back to that relative level. The estimated percentage for 1982 is only 12.07. Although there appears to have been a slight reversal in this trend during the late 1970s (for example, 1979), the estimated 1982 percentage is the lowest in nearly two decades.

TABLE 3.3
Basic Research as a Percentage of Total R&D in the U.S. Economy

Years	*Percentages*
1953	8.02
1965	12.75
1971	13.77
1976	12.78
1979	13.20
1980 (prelim)	12.97
1981 (est.)	12.61
1982 (est.)	12.07

Source: Calculated from National Science Foundation, *Research and Development in Industry, 1980* (U.S. Government Printing Office, 1982), pp. 28-29.

Industry Trends

A similar pattern of decline in total R&D, and a reallocation of R&D away from basic research also occurred at the industry level. The percentage of industrial R&D allocated to basic research peaked at 4.17 percent around 1965 and has been declining thereafter. These percentages are shown in Table 3.4. In comparing the data in Tables 3.3 and 3.4, the percentage of R&D allocated to basic research at the national level peaked nearly six years after the decline in the percentage of R&D allocated to basic research within the industrial sector. This lagged peak in the national R&D statistics is due to the continued federal support of basic research (military and space oriented) during the late 1960s.

TABLE 3.4
Basic Research as a Percentage of Total R&D in the Industrial Sector

Years	*Percentages*
1953	4.16
1965	4.17
1970	3.33
1975	3.02
1979	3.02

Source: Calculated from National Science Foundation, 1982b, p. 40.

The National Science Foundation's analysis of the basic/applied/development mix of R&D spending is the most comprehensive source of published industrial data on the composition of R&D (in total or by specific industry). However, other categories of R&D activity are perhaps equally as meaningful. To our knowledge, the only other published source of disaggregated R&D is the McGraw-Hill (1981, 1982) industry breakdown of R&D into the three categories: R&D allocated to (1) new products, (2) new processes, and (3) improving existing products. For 1981, McGraw-Hill reported that the majority of R&D in the manufacturing sector went toward product related activities: 40 percent to new products, 17 percent to new processes, and 43 percent to improving existing products.

Mansfield (1981a) has recently reported industry summary statistics for three other important classifications. He found, from a survey of 108 manufacturing firms, that in 1977 an average of 32 percent of company-financed R&D went to "projects lasting five or more years" (referred to as long-term R&D), 33 percent to "projects aimed at entirely new products and processes," and 23 percent to "projects with less than a 50-50 estimated chance of success" (referred to as risky R&D).

Clearly, there are many meaningful ways to categorize industrial R&D spending. R&D is not homogeneous in its nature.

As part of our 1982 survey (discussed in Chapter 2) we obtained data from 146 industrial manufacturing firms on the disaggregated nature of their R&D programs and on how their allocations of R&D have been changing during the past decade. Specifically, we obtained each firm's estimate of the percentage of its R&D allocated to (1) basic research, as opposed to applied research or development, (2) long-term research ("projects estimated to last longer than five years"), and (3) either entirely new production processes, improvements in existing production processes, entirely new products, or improvements in existing products. The sum of the four percentages in category (3) equals 1.0.[1]

Our survey findings are summarized in Table 3.5. The pattern of behavior regarding basic research expenditures has some similarities to that for the entire industrial sector, as shown in Table 3.4. The reported survey percentages in column (1) of Table 3.5 show the 1977 and 1978 estimates to be somewhat lower than for 1970; however, there appears to be an expected increase in basic research outlays over the next decade. Although we are skeptical about the validity of the exact value of the predicted 1985 and 1990 estimated percentages for basic research, the direction of the indicated trend probably reflects current expectations. This anticipated change away from development toward basic research is consistent with the expected increase in R&D outlays in the 1980s toward the long-term, as opposed to short-term, end of the continuum. As seen from column (2), the percentages of R&D going to long-term projects fell during the 1970s but is predicted to be on the increase. The numbers in columns (3) through (6) reinforce our earlier statements about the heterogeneity of industrial R&D. Clearly, the lion's share of industrial R&D during the last decade has been directed toward the improvement of existing process and product technology, and there does not appear to be evidence that a significant reallocation is likely by 1990. The mean percentages do show a slight increase in R&D aimed at new process and new product innovations beginning in 1980. However, the percentage increases are not as striking as are the percentage increases for either basic research or for long-term activities.

A new category of industrial R&D spending became important in the mid-1960s as the effect of environmental regulations on production costs began to rise. The need for environmental regulations, pollution control in particular, began to receive congressional attention shortly after World War II – the 1948 Water Pollution Control Act and its 1956 amendments, the 1955 Air Pollution Control Act, and the 1963

TABLE 3.5
Mean Percentages of Company-Financed R&D Allocated to Basic Research, Long-Term Projects, and Product or Process Activities

Years	(1) *Basic*	(2) *Long-Term*	(3) *New Production Processes*	(4) *Improvements in Existing Production Processes*	(5) *New Products*	(6) *Improvements in Existing Products*
1970	2.84%	23.10%	15.47%	21.44%	25.53%	40.76%
1977	2.79	22.39	15.59	22.47	24.38	40.71
1978	2.78	22.64	15.11	22.81	25.81	39.78
1979	3.22	22.95	15.02	22.27	26.15	39.47
1980	3.51	22.26	14.69	23.46	28.03	38.66
1981	3.83	23.75	15.85	23.08	29.29	37.47
1985 (est.)	4.72	24.09	16.17	22.54	29.31	36.14
1990 (est.)	5.16	24.54	16.53	20.90	30.22	35.53

Source: 1982 survey of 146 firms by authors.

Clean Air Act – but this early legislation did not impose direct costs on industry. More recent regulations have imposed direct costs. Some of the more prominent statutes include the 1965 Water Quality Act, the 1965 Motor Vehicle Air Pollution Control Act, the 1967 Air Quality Act, the 1970 Clean Air Amendments, and the 1972 Water Pollution Control Act Amendments. In general, the effect of environmental regulations on production costs can be ignored before 1967. Since 1967, however, environmental compliance has caused the capital-related costs of production to rise significantly. The Industrial Research Institute estimates that R&D expenditures allocated by all industry for environmental improvements grew at an average annual rate of 15.4 percent between 1974 and 1977 (Manners and Nason 1978). The National Science Foundation (1982a) reports that between 1973 and 1979 company-financed R&D allocated for industrial pollution abatement grew at an average annual rate of 16.8 percent. Over that time period industrial R&D, as a total, grew at only 9.4 percent. Thus, R&D expenditures allocated toward environmental compliance may be a factor in the shift of industrial R&D away from the basic, long-term ends of the R&D spectrum. (As we discuss in Chapter 5, these redirected R&D dollars may be one determinate of the productivity slow-down in the United States since mid-1960).

To our knowledge the only detailed published estimates of industrial trends in R&D allocated toward environmental compliance are in Link (1982f). Based on survey data for 97 major R&D firms in the chemicals (32 firms), machinery (51 firms), and petroleum (14 firms) industries, Link found that in 1977 the mean of company-financed R&D allocated to comply with environmentally related government regulations was 8.9 percent. As reported in column (1) of Table 3.6, the mean percentages vary across the three industries studied; 4.8 percent in machinery, 9.7 percent in chemicals, and 23.7 percent in the petroleum industry. The average annual rate of growth in these compliance related expenditures varied by industry. As reported in column (2) the firms responding from the chemicals and the petroleum industries indicated a greater than 60 percent average annual rate of growth in compliance related R&D between the years 1977 and 1980. Associated with these increases, the firms also noted that R&D funds allocated to meet environmental regulations were taken at the expense of applied research and development spending, rather than at the expense of basic research. The percentage of compliance related R&D that would have gone to basic research was only 4.1 percent in chemicals, 1.6 percent in machinery, and 0.8 percent in the petroleum industry.

TABLE 3.6
R&D Allocated to Comply with Environmentally Related Government Regulations

Industry	*(1)* *1977 Mean Percentage*	*(2)* *Annual Growth Rate Between 1977-80*
Chemicals	9.7%	61.5%
Machinery	4.8	23.2
Petroleum	23.7	70.7

Source: 1982 survey of 146 firms by authors.

CONCLUSIONS

The data presented in this chapter indicate clearly that R&D is quite heterogeneous in scope. Industrial firms do not think in terms of R&D as one composite form of innovation-related investments. Rather, there are many separate categories that have meaning. The terms basic/applied/development appear well known, not so much because R&D activities neatly fall into these groupings, but because most firms are familiar with the external reporting of R&D (which uses these National Science Foundation definitions). The long-term versus short term and product versus process dichotomies are more closely aligned with industrial experience.

Since the mid-1960s it appears that R&D spending has been increasingly allocated to the more short-term applied types of activities. The logical question to ask is: Why has this reallocation been taking place? We attempt to answer this question in the next chapter. There, we first consider why firms differ in their R&D mix and then we consider why the composition of firms' R&D has been changing at differing rates.

NOTES

1. As part of the 1978 survey, we obtained data for 1977 on the allocation of each firm's R&D to either (1) basic, applied, or development activities, or (2) activities leading to product versus process innovations. These data will be investigated in detail in Chapter 4.

4

An Analysis of the Composition of R&D Spending

INTRODUCTION

As we discussed in Chapter 2, firms differ in the extent to which they use indigenous R&D spending as a source for acquiring innovation-related technical knowledge. Firms also differ considerably in the extent to which they devote financial resources to R&D. The data in Table 4.1 indicate that company-financed R&D as a percent of net sales varies across industries within the manufacturing sector. The range is from 5.6 percent in the professional and scientific instruments industry to 0.5 percent in the petroleum industry. Within any one industry there is also significant inter-firm variation in R&D activity. Among the R&D firms in the chemicals industry, for example, R&D spending in 1981 ranged from $631 million (DuPont) to $1.1 million (Mississippi Chemical), and R&D spending as a percent of sales in 1981 ranged from 5.1 percent (Hunt Chemical) to 0.3 percent (Ferro) (Business Week 1982).[1] Similar patterns exist in all industries.

There have been numerous empirical studies of inter-firm differences in R&D spending.[2] The purpose of these studies is to define the environment (that is, the set of industry and firm characteristics) most conducive for innovative activity. These studies have been useful.

Earlier versions of the materials presented in this chapter were first published in the *Journal of Industrial Economics* (Link and Long 1981), *IEEE Transactions on Engineering Management* (Higgins and Link 1981), *The Transfer and Utilization of Technical Knowledge* (Link 1982a), and the *Southern Economic Journal* (Link 1982b). Portions of all of these research projects were funded by a grant from the National Science Foundation, Division of Policy Research and Analysis (PRA 8009552).

TABLE 4.1
Company-Financed R&D Spending as a Percent of Net Sales: 1980

Industry	*Percentage*
Lumber, wood products and furniture (SIC 24, 25)	0.8
Chemicals (SIC 28)	3.2
Petroleum refining (SIC 29)	0.5
Primary metals (SIC 33)	0.6
Fabricated metals (SIC 34)	1.2
Machinery (SIC 35)	4.8
Electrical equipment (SIC 36)	3.9
Motor vehicles (SIC 371)	4.3
Aircraft and missiles (SIC 372, 376)	3.3
Professional and scientific instruments (SIC 38)	5.6
All Industries	2.2

Note: SIC – Standard Industrial Classification.

Source: National Science Foundation. *Research and Development in Industry*, 1980 (U.S. Government Printing Office, 1982), p. 24.

From them a generally accepted set of variables has been found that are correlated with inter-firm levels of R&D spending. However, virtually nothing is known about the determinants of inter-firm differences in the composition of R&D. As we maintained in Chapter 3, R&D is heterogeneous in nature and a more complete understanding of the role of R&D in innovation can only be obtained when the composition of R&D is more fully considered.

The purpose of this chapter is to present some empirical findings on the determinants of inter-firm differences in R&D expenditures disaggregated by character of use. Two disaggregation schemes are considered. The first is the basic/applied/development trichotomy and the second is a product/process breakdown.

AN ANALYTICAL FRAMEWORK OF THE DETERMINANTS OF R&D

From the existing empirical studies of inter-firm differences in R&D spending, six variables are generally analyzed. Five of the variables are firm characteristics and one is a characteristic of the industry in which the firm operates. These determinants include the firm's size, its profitability, the extent of the firm's diversification, the amount of federal R&D received by the firm, the firm's form of ownership control, and the market structure of the industry in which the firm sells its product. The motivation for each of these variables is well founded in the literature and is only briefly reviewed below.

Firm size. Since Schumpeter (1947) wrote *Capitalism, Socialism and Democracy*, economists have been willing to associate economic growth through innovation with, among other things, large firm size. This aspect of the so-called Schumpeterian hypothesis is posited on the view that in a capitalistic system economic growth occurs through the process of "creative destruction" whereby the "old" industrial structure – its products, its processes, or its organization – is continually changed by "new" industrial innovations. The motivating force behind this process is the promise of economic profit achieved through innovative activity. According to Schumpeter, large firm size is essential to the success of such innovative activity. Simply, larger firms can provide economies of scale in production and in innovation that make available sufficient resources necessary for successful completion of the process.

During the past two decades there have been an abundant number of statistical tests of this aspect of Schumpeter. A positive relationship between innovative activity, generally measured in terms of R&D related activities, and some measure of firm size is sought (Schmookler 1952; Villard 1958; Worley 1961; Horowitz 1962; Hamberg 1964, 1966; Scherer 1965a, 1965b, 1967a; Comanor 1967; Mueller 1967; Mansfield 1968; Adams 1970; Phlips 1971; Loeb and Lin 1977; Shrieves 1978; Soete 1979; Armour and Teece 1980; Link 1980, 1981c). Kamien and Schwartz (1982) have perhaps best summarized the statistical evidence from these studies as "wanting" in their support of Schumpeter.

There are several reasons for this lack of empirical verification. Grabowski and Mueller (1970) suggest that the industrial complexes of today have changed the structure of capitalism envisioned by Schumpeter. An alternative explanation posited by Markham (1965) and formalized by Fisher and Temin (1973) is that the empirical literature relating R&D to firm size is not a justifiable test of Schumpeter. More recently, Lunn (1982) has demonstrated theoretically that the relationship between innovative inputs and firm size cannot be used to infer the relationship between innovative output and firm size. Also, according to Hamberg (1966) the casual tendency to identify large industrial laboratories (which are generally connected with the larger-sized firms) with major innovative activities (which are presumed to be R&D intensive) is often a mistake. Historically, most major inventions and innovations have not come from the major R&D labs. Even if the empirical models used by the researchers cited above were an accurate interpretation of Schumpeter (and we agree that they are not), testing the hypothesis with a sample of only the largest industrial firms does not accurately measure the total influence of varying firm sizes.

Firm profits. Innovation, like any future-oriented activity, involves risk and uncertainty (Gilpin 1975; Nelson and Winter 1977). Such activities require substantial, and often prolonged, financial support. External funding is generally difficult or costly to obtain for R&D; thus internal sources of financing must be available. For empirical purposes these sources are generally assumed to be related to current or past profitability.

There are two interpretations of the role of profits as a determinant of R&D. First, profits are an expectational proxy. Large or growing profits indicate business success and thus encourage management to undertake additional R&D (Horowitz 1961; Mueller 1967; Elliott 1971). Second, large or growing profits indicate the availability of internal funds, liquidity, for the financing of R&D (Meyer and Kuh 1957; Grabowski 1968). With either interpretation,[3] previous researchers have hypothesized and have found some support for a positive relationship between R&D expenditures and profitability (Horowitz 1961; Mueller 1967; Grabowski 1968; Wilson 1977; Armour and Teece 1980). Studies often cited as failing to support the R&D to profitability relationship have not measured R&D activity in terms of dollar expenditures, but rather have examined either R&D related employment figures or patent statistics (Scherer 1965a; Hamberg 1966; Johannisson and Lindstrom 1971; Smyth, Samuels, and Tzoannos 1972; Branch 1974; McEachern and Romeo 1978).

Kamien and Schwartz (1978) suggest that the R&D-to-profits relationship is stronger in some firms than in others. Their theoretical analysis concludes that marginal firms, or those engaged in "risky" R&D ventures, face an active self-financing constraint and thus high current profits are a necessary condition for R&D spending. Established firms doing "routine" product development will not be constrained by cash availability and will not require high current profits to finance their R&D. Therefore, although the motivation for a profits variable is well documented, the precise direction of influence remains an open question.

Firm diversification. Nelson's (1959) pioneering study of the economics of basic scientific research stressed that the results of such research are often unpredictable. It is not unusual for technical knowledge to simply sit, or to be used in the future as an additional research input: basic research is not intended to solve the immediate practical problems of the firm. Because of this characteristic, Nelson hypothesized that only diversified firms with broad technological bases, and with "their fingers in many pies" (1959, p. 302), will find it profitable to engage in basic research, other things held constant, since they are better able to market whatever inventions or discoveries result.

Subsequent researchers (Comanor 1965; Scherer 1965a; Grabowski 1968; McEachern and Romeo 1978) have relied on Nelson's so-called diversification hypothesis to formulate empirical inquiries into the role of product diversification as a determinant of R&D activity. Although these investigators are cognizant of the fact that Nelson's arguments are specific to basic research, they still seek a positive relationship between total R&D and a measure of product diversification. Grabowski (1968) avers that so long as basic research is not a perfect substitute for applied research or development, Nelson's argument is generally applicable. The empirical evidence in these studies is, not surprisingly, mixed. Direct tests of Nelson's hypothesis by Link (1981a, 1982b) and by Link and Long (1981) support it.

Federal R&D conducted by the firm. Very little is known about the relationship between private and federal R&D in either the innovation process or the production of technology. Some speculations do however exist, the origin of which can be traced to the "pump-priming" versus the "substitution" hypotheses of Blank and Stigler (1957). On the one hand, Hamberg (1966) claims that new scientific knowledge resulting from federally sponsored R&D projects enlarges the firm's scientific base and thus expands its opportunity set for additional private R&D expenditures. Tilton (1973) and Kendrick (1978), on the other hand, suggest that federal R&D and private R&D may be substitutes in the firm's innovation process. Increases in federal allocations may displace private investments if (1) the resulting R&D output can be internalized by the firm, or if (2) federal obligations cause the firm to reach its capacity for technical operations. The strength of these potentially offsetting effects of federal R&D on private R&D, or more broadly on private innovative activity, is an empirical question. Shrieves's (1978) evidence suggests a crowding-out of private R&D by federal R&D; however, the studies by Leonard (1971), Goldberg (1978), Higgins and Link (1981), Levy and Terleckyj (1983), Link (1982a, 1982b, 1982d), Mansfield and Switzer (1982) and Scott (1981) can be interpreted as supporting an argument in favor of complementarity.

Ownership form of control of the firm. In traditional neoclassical theory the owner of the firm (the risk bearer) and the manager of the firm (the risk taker) are the same person. However, in the more modern managerial theories of the firm, as popularized by Berle and Means (1932), there is a delineation of behavioral characteristics between owner-controlled and manager-controlled firms. One particular difference relates to the firm's attitude toward risk.

According to the managerial theories, manager-controlled firms are relatively more risk averse than owner-controlled firms because of an asymmetry in the manager's reward structure. Simply, hired managers are more risk averse than owners since they do not internalize the benefits from successful risk taking but do internalize the costs from unsuccessful risk taking activities. The hired manager benefits from successful risk-related ventures only to the extent that he is a stockholder and that successful ventures result in higher profits and stock prices: unsuccessful ventures may result in the manager's losing his job. This risk aversion hypothesis is often attributed to Monsen and Downs (1965) and Baumol (1967); however, its origins can be traced at least to Beard (1938) and Gordon (1945). Beard wrote:

> Tied down by fiduciary responsibility to a great structure, forced to consider the stockholder, ... as well as the board of directors, the position of [the manager] was conducive to anything but "initiative" and "leadership." Caution became his watchword [and] he shunned ... risky new enterprises. (1938, p. 727)

Along a similar line of thought, Gordon wrote:

> The "institutional environment" of the salaried executive [manager] ... differs from that of the owner-entrepreneur .... [A]mong some ... executives, scientific caution may degenerate into a tendency to play safe. [Managers] do not receive the profits which may result from taking a chance, while their position in the firm may be jeopardized in the event of serious loss. (1945, pp. 323-24)

There have been many empirical studies of this manager-related risk aversion hypothesis as it relates to general corporate behavior; but the findings are not conclusive (Larner 1970; Boudreaux 1973, 1975; Bower 1974; Holl 1975; McEachern 1975; Kania and McKean 1976; Qualls 1976; Stano 1976; McEachern and Romeo 1978).

For the manager-related risk aversion hypothesis to apply to R&D activity, the nature of the risks involved in the R&D must be specified clearly. There are at least two dimensions of risk to consider.

There is risk in undertaking too much R&D and failing, and there is risk in not undertaking enough R&D and thus losing market shares to the more innovative firms. The effect of ownership control on R&D per se is unknown and thus remains an empirical question. The research of McEachern and Romeo (1978) suggests that firms controlled by individuals outside of direct management undertake a defensive innovation strategy and are thus relatively more R&D intensive. Their findings are preliminary.

Industry concentration. Schumpeter (1947) not only stressed the importance of firm size as a prerequisite for innovation, but also suggested the monopoly power is equally as important — an obvious contradiction to the Smithian view of monopoly's control. Within the process of "creative destruction," firms with monopoly power will find innovative activity attractive.

More recent theoretical analyses have concerned themselves with this issue. In particular, Arrow has analyzed the incentives offered the firm to invent under various market structures and has concluded that "the incentive to invent is less under monopolistic than under competitive conditions" (Arrow 1962, p. 619). Demsetz has criticized Arrow's analysis on the grounds that it ignored the effect of industry size on inventive actions. This theoretical reconsideration concluded that "the incentive to [invent] is just the reverse of what Arrow concluded" (Demsetz 1969, p. 17). There have been several theoretical attempts to reconcile this Arrow-Demsetz controversy (Kamien and Schwartz 1970; Hu 1973); however, the issue remains unresolved.

These theoretical models explicitly assume that innovation and inventive activity are cost reducing, or process augmenting. However, most so-called tests of this aspect of Schumpeter have simply been directed toward correlating industry concentration and R&D spending. In fact, these tests are really based on Villard's (1958) corollary to Schumpeter that firms in concentrated industries do relatively more R&D (Brozen 1965).

Nevertheless, a few researchers have found statistical support for the proposition that firms in concentrated industries do relatively more R&D. Most agree that the relationship between market structure and R&D is complex, and some agree that the relationship is simultaneous (Dasgupta and Stiglitz 1980a, 1980b; Levin and Reiss 1982). Others simply conclude that there are more important variables, as those discussed above, driving the R&D effort of the firm (Scherer 1965a, 1967a; Phillips 1966; Comanor 1967; Globerman 1973; Rosenberg 1976; Wilson 1977; Shrieves 1978).

On the basis of the arguments associated with these six variables, researchers have considered regression models where the firm's R&D intensity (RD/SIZE) is the dependent variable and the firm's relative profits (PROF/SIZE), its extent of product diversification (DIV), its receipt of federal R&D (FRD/SIZE), its form of ownership control (OWNER), and its industry concentration (CONC) are the independent variables:

$$(4.1) \qquad \text{RD/SIZE} = \alpha_0 + \alpha_1 \,\text{PROF/SIZE} + \alpha_2 \,\text{DIV} + \alpha_3 \,\text{FRD/SIZE} + \alpha_4 \,\text{OWNER} + \alpha_5 \,\text{CONC} + \varepsilon,$$

where ε is an error term assumed to obey all of the classical assumptions. On the basis of the arguments summarized above, $\hat{\alpha}_1$, $\hat{\alpha}_2$, and $\hat{\alpha}_5$ are expected to be greater than zero. There is also evidence in the literature to expect $\hat{\alpha}_3 > 0$, but no prediction is made regarding the sign of $\hat{\alpha}_4$. These four testable propositions will be examined below.

We are also concerned in this chapter with the influence of each of these independent variables on the composition of R&D, that is, on the percentage of the firm's R&D allocated to either basic research (BR), applied research (AR), and development (D) or to product innovation (PROD) and process innovation (PROC). Specifically, the following models are of interest:

$$(4.2) \qquad \text{BR} = \beta_0 + \beta_1 \,\text{PROF/SIZE} + \beta_2 \,\text{DIV} + \beta_3 \,\text{FRD/SIZE} + \beta_4 \,\text{OWNER} + \beta_5 \,\text{CONC} + \varepsilon_1,$$

$$(4.3) \qquad \text{AR} = \gamma_0 + \gamma_1 \,\text{PROF/SIZE} + \gamma_2 \,\text{DIV} + \gamma_3 \,\text{FRD/SIZE} + \gamma_4 \,\text{OWNER} + \gamma_5 \,\text{CONC} + \varepsilon_2,$$

$$(4.4) \qquad \text{D} = \eta_0 + \eta_1 \,\text{PROF/SIZE} + \eta_2 \,\text{DIV} + \eta_3 \,\text{FRD/SIZE} + \eta_4 \,\text{OWNER} + \eta_5 \,\text{CONC} + \varepsilon_3,$$

$$(4.5) \qquad \text{PROD} = \delta_0 + \delta_1 \,\text{PROF/SIZE} + \delta_2 \,\text{DIV} + \delta_3 \,\text{FRD/SIZE} + \delta_4 \,\text{OWNER} + \delta_5 \,\text{CONC} + \varepsilon_4,$$

and,

$$(4.6) \qquad \text{PROC} = \varrho_0 + \varrho_1 \,\text{PROF/SIZE} + \varrho_2 \,\text{DIV} + \varrho_3 \,\text{FRD/SIZE} + \varrho_4 \,\text{OWNER} + \varrho_5 \,\text{CONC} + \varepsilon_5,$$

where the error terms ε_i are assumed to obey all of the classical assumptions.

A matrix is presented as Table 4.2 to summarize testable propositions associated with the various categories of R&D spending. The testable propositions associated with RD/SIZE and the independent variable have been discussed above and are summarized again in Table 4.2. We also noted above that the theoretical work of Kamien and Schwartz (1978) concludes that high current profits are a necessary condition for "risky" R&D ventures. We do not consider a separate category of R&D spending that dichtomizes R&D into risky and nonrisky activities; although such an experiment would be useful. There is risk associated with each of the five categories of R&D considered; however, it is reasonable to expect that there is more uncertainty (by definition) associated with basic research than with either applied research or development (Manners and Nason 1978). Therefore, following Kamien and Schwartz, one testable proposition is that there is a positive relationship between BR and PROF/SIZE, that is, $\hat{\beta}_1 > 0$.

Nelson's (1959) diversification hypothesis was posited specifically with reference to basic research. Therefore, our second testable proposition is that DIV and BR are also positively correlated, that is, $\hat{\beta}_2 > 0$.

The empirical evidence available suggests that federal R&D allocations are positively related to the total amount of R&D financed by the firm. No arguments have been made as to what categories of R&D are increased in response to an increase in federal R&D. Most federally contracted research is product oriented. One may thus argue that the receipt of such a contract will change the marginal cost of a firm's own product-related innovative activities. On the one hand, if the firm can internalize the cost-related benefits from increasing its scale of total development activities or if the firm can transfer knowledge embodied in the labor assigned to the contracted project to other company specific projects, then budgeted company funds for product-related activities may be displaced. On the other hand, if decreasing returns to scale in product innovation set in as a result of acquiring federal contracts, then the firm may increase its product-related allocations. Either case is, of course, possible under a given set of conditions. Therefore, the net impact of federal R&D on the firm's allocation of its own R&D between alternative uses remains an empirical question.

The modern managerial theories of the firm delineate behavioral characteristics in relationship to the firm's attitude toward risk. All other things remaining constant, manager-controlled firms are presumed to be relatively more risk averse than owner-controlled firms. As we argued above, there is risk associated with all dimensions of R&D, but it is generally believed that there is more uncertainty associated with basic research. It follows, therefore, from the managerial theories that owner-controlled firms may be less averse to engaging in basic research than manager-controlled firms. If OWNER is interpreted as an index of owner-controlled firms, then a third testable proposition is that there is a positive relationship between BR and OWNER, that is, $\hat{\beta}_4 > 0$.

Although much has been written about the relationship between market power and innovative activity, very few conclusions can be supported empirically. The traditional attempts to correlate a concentration measure and total R&D spending are clearly an oversimplification of the issue. One possible approach toward a meaningful reconciliation is to examine the relationship between market power and particular categories of R&D spending. As previously discussed, the central Schumpeterian idea is that market power impacts on cost-reducing, or process-augmenting, innovations; not all R&D is directed toward such a goal (as was noted in Chapter 3). A related body of literature focuses on the influence of market structure on the firm's incentive to increase the pace at which it introduces new innovations. The theoretical arguments posited by Scherer (1967b) and Kamien and Schwartz (1976) suggest that less-concentrated industries are more likely to be characterized by competition from rivals for market shares. This rivalry, in turn, increases the pace of new innovations — product-related innovations in particular (Loury 1979; Lee and Wilde 1980) — since firms may have much to lose from being the second innovator. Thus, a fourth testable proposition is that there is a positive (negative) relationship between CONC and PROC (PROD) oriented R&D, and a negative relationship between CONC and D. That is, $\hat{\varrho}_5 > 0$, $\hat{\delta}_5 < 0$, and $\hat{\eta}_5 < 0$. In the following section each of these propositions is tested empirically.

THE EMPIRICAL ANALYSIS

Equations (4.1) through (4.6) are estimated using data for 275 of the 329 manufacturing firms in the 1978 survey. To obtain an indication of the representativeness of the sample, the percentage allocations from the survey sample were calculated and compared to published industry statistics. Regarding the basic/applied/development breakdown, the mean responses from the survey sample are 3.8 percent to basic research, 25.2 percent to applied research, and 69.5 percent to development. These figures are quite similar to the National Science Foundation (1982b) aggregates for all manufacturing; 3.6 percent to basic research, 22.1 percent to applied research, and 74.3 percent to development. Industry-by-industry comparisons of these

TABLE 4.2
The Predicted Direction of Influence of Firm and Industry Characteristics on R&D Spending

Independent Variables	*Dependent Variables*					
	RD/SIZE	*BR*	*AR*	*D*	*PROD*	*PROC*
PROF/SIZE	+	+	?	?	?	?
DIV	+	+	?	?	?	?
FRD/SIZE	+	?	?	?	?	?
OWNER	?	+	?	?	?	?
CONC	+	?	?	-	-	+

Source: Compiled by the authors.

percentages are presented in Table 4.3 for those selected industries for which published aggregate statistics are available.

Similarly, the mean sample percentage responses for the allocation of R&D between product and process innovations were also computed. There is no published set of data that precisely corresponds to the product versus process survey breakdown; however, the McGraw-Hill (1978) statistics may serve as a crude measure of our sample's representativeness. For the sample, the mean percentage of R&D going to product innovation is 70.3 percent and to process innovation is 29.7 percent. McGraw-Hill (1978) reports for the entire manufacturing sector 83 percent product and 17 percent process. Industry-by-industry comparisons are presented in Table 4.4.

There are two fundamental questions about the validity of these firm specific survey responses that need to be addressed. The first question relates to the general appropriateness of the categories basic research, applied research, and development. Since the 1950s the National Science Foundation has used these labels in an attempt to compare R&D expenditures in all sectors of the economy using a common set of definitions. Nason, Steger, and Manners (1978) have considered, in particular, the appropriateness of the label basic research for survey purposes and have concluded that the label does have construct validity. Although most firms categorize R&D differently they are, according to Nason, Steger, and Manners, familiar enough with National Science Foundation reporting that the terms are meaningful. Presumably, this would also be the case for the terms applied research and development.

TABLE 4.3
Comparison of the Allocation of R&D: Basic/Applied/Development

Industry	*Survey Sample*			*Manufacturing Sector*		
	Mean % Basic	*Mean % Applied*	*Mean % Development*	*% Basic*	*% Applied*	*% Development*
Lumber and furniture (SIC 24, 25)	2.8	39.0	58.5	5.5	29.1	65.4
Chemicals (SIC 28)	7.8	37.4	54.8	9.0	41.9	49.1
Fabricated products (SIC 34)	1.7	30.7	67.5	0.6	24.7	74.7
Machinery (SIC 35)	2.9	17.5	79.5	1.7	9.0	89.3
Electrical equipment (SIC 36)	2.0	18.8	79.2	4.8	21.0	74.3
Professional & scientific instruments (SIC 38)	1.2	16.7	82.1	1.5	13.8	84.7
Manufacturing Sector	3.8	25.2	69.5	3.6	22.1	74.3

Note: SIC – Standard Industrial Classification.

Source: 1978 survey by authors. National Science Foundation, *Research and Development in Industry, 1980*, (U.S. Government Printing Office, 1982).

TABLE 4.4
Comparison of the Allocation of R&D: Product and Process Innovations

Industry	Survey Sample: Mean % Product	Survey Sample: Mean % Process	Manufacturing Sector: % Product	Manufacturing Sector: % Process
Food and kindred products (SIC 20)	59.1	40.9	67.0	33.0
Textiles & apparel (SIC 22, 23)	54.2	45.8	74.0	26.0
Lumber & furniture (SIC 24, 25)	70.1	29.9	90.0	10.0
Paper & products (SIC 26)	57.0	43.0	70.0	30.0
Chemicals (SIC 28)	66.4	33.6	80.0	20.0
Petroleum refining (SIC 29)	48.9	51.1	75.0	25.0
Rubber & plastic products (SIC 30)	66.4	33.6	82.0	18.0
Stone, clay, glass (SIC 32)	45.9	54.1	77.0	23.0
Primary metals (SIC 33)	49.7	51.3	62.0	38.0
Fabricated metal products (SIC 34)	68.4	31.6	72.0	28.0
Machinery (SIC 35)	80.2	19.8	86.0	14.0
Electrical equipment (SIC 36)	70.9	29.1	94.0	6.0
Transportation equipment (SIC 37)	61.9	38.1	75.0	25.0
Professional & scientific instruments (SIC 38)	83.1	16.9	87.0	13.0
Manufacturing Sector	70.3	29.7	83.0	17.0

Note: SIC — Standard Industrial Classification.

Source: 1978 survey by authors. "23rd Annual McGraw-Hill Survey of Business' Plans for Research and Development, 1978-81," (Mimeographed, 1978).

The second question relates to the product and process innovation labels. There is, over time, a symbiotic relationship between the firm's product- and process-related activities. This fact was illustrated in the theoretical analysis of Chapter 2. In the chemicals industry, for example, process innovation is often the first stage in the development of a new product. Here, however, these two categories are viewed, at a given point in time, as alternative investment choices. In addition, this product/process dichotomization conforms well to economic theory. For example, process-related innovations directly relate to a conceptual shift of a production isoquant. More specifically, economic theories of inventive behavior are implicitly based on process-related changes.

The Data

Data on the dependent variables in equations (4.1) through (4.6) come from the 1978 survey of 329 manufacturing firms. For each firm the sum of the percentages of self-financed R&D going to either basic research/applied research/development is 100 percent, as is that going to product innovation/process innovation. That is, for each firm, (BR + AR + D = 1.0) and (PROD + PROC = 1.0).

For equation (4.1), data on total 1977 expenditures on R&D come from Compustat. The denominator, firm size, represents the firm's net sales defined as gross sales and other operating revenue less discounts, returns, and allowances, as reported by Compustat. Firm profits, measured as profits after taxes and dividends, lagged one year, are also reported by Compustat. Data on federal R&D expenditures received by each firm were obtained as part of the 1978 survey. Each of these variables is measured in millions of dollars.

The extent of product diversification characterizing each firm is represented by the number of four-digit SIC industries in which the firm and its subsidiaries operate. These data come from the *Register of Corporations* and were used in the analysis in Chapter 2.

The index of ownership control used here is formulated following the work of McEachern (1975). Each firm is classified into one of three categories: (1) externally controlled — EC equals one if a party holds 4 percent or more of the common stock but is not a part of management, and EC equals zero otherwise; (2) manager controlled — MC equals one if no single party owns 4 percent or more of the common stock, and MC equals zero otherwise; and (3) owner managed — OM equals one if the chief executive officer directly or beneficially owns 4 percent or more of the common stock, and OM equals zero otherwise. In this trichotomy the owner-manager is presumed to be the least risk averse coming closest to a classical entrepreneur: the externally controlled firm is presumed to be the most risk averse since an outside dominant stockholder is believed to have little knowledge of the firm's daily activities and hence may be myopic in judgement.

Finally, a four-firm concentration ratio, corresponding to the four-digit SIC industry in which each firm produces, approximates the market power of the firms. These data come from the *1972 Census of Manufactures*.

The Statistical Results

Equation (4.1) and equations (4.2) through (4.6) are estimated using ordinary least-squares analysis. Since (BR + AR + D = 1.0) and (PROD + PROC = 1.0), it follows that the error terms are also related across equations as ($\varepsilon_1 + \varepsilon_2 + \varepsilon_3 = 0$) and ($\varepsilon_4 + \varepsilon_5 = 0$). However, since the independent variables are the same in each equation, no efficiency is gained by using another estimation technique (such as seemingly unrelated regression estimation).

The least-squares results, with t-statistics in parentheses, corresponding to equations (4.1) through (4.4) are presented in Table 4.5. The results reported in column (1) correspond very well to those reported by others. The estimated coefficients on the profit and federal R&D variables are significantly related to total R&D intensity. The latter finding supports the argument for complementarity between private and federal R&D: a one dollar increase in federally financed R&D increases company-financed R&D by 9.4 cents. The estimated coefficient on the diversification variable is also positive, but only marginally significant. Finally, the ownership and concentration variables are not statistically important.

The least-squares results from equations (4.2) through (4.4) are also reported in columns (2) through (4) of Table 4.5. As predicted, profitability is positively related to the percentage of R&D going to basic research. This finding is in direct support of the Kamien-Schwartz and Nelson propositions discussed above. Also, the estimated coefficient on CONC is negative and marginally significant. Although we did not make such a prediction, this finding agrees with those reported by Mansfield (1981a). Whereas the results in column (1) suggest that an increase in federal R&D increases private R&D, the results in column (2) suggest that this increase in private R&D is primarily directed away from basic research and toward development. A similar conclusion was reported by Higgins and Link (1981). Finally, the estimated coefficient on the owner-manager variable is positive and significant. Whereas owner-manager firms are no more R&D intensive than firms with other forms of ownership control, other things

TABLE 4.5
Least-Squares Regression Results
from Equations (4.1)-(4.4) (t-statistics in parentheses)

	Dependent Variables			
Independent	*(1)*	*(2)*	*(3)*	*(4)*
Variables	*RD/SIZE*	*BR*	*AR*	*D*
Intercept	0.027	0.007	0.256	0.728
	(3.76)**	(1.21)	(1.41)	(2.01)*
PROF/SIZE	0.258	0.153	0.695	-0.848
	(2.99)**	(3.06)**	(2.26)*	(-1.13)
DIV	0.0038	0.0022	0.0009	-0.0031
	(1.72)	(2.27)*	(1.05)	(-1.37)
FRD/SIZE	0.094	-0.081	-0.005	0.086
	(2.01)*	(-1.75)	(-1.59)	(2.31)*
MC	-0.0003	0.009	0.011	-0.02
	(-0.72)	(0.43)	(0.29)	(-0.45)
OM	0.0007	0.004	-0.001	-0.003
	(0.41)	(2.14)*	(-0.51)	(-1.34)
CONC	0.017	-0.0041	-0.0007	0.0048
	(0.87)	(-1.71)	(-0.91)	(2.15)*
R^2	0.385	0.462	0.317	0.401

** significant at the 0.01 level.
* significant at the 0.05 level.
Source: Albert N. Link, "An Analysis of the Composition of R&D Spending." *Southern Economic Journal* 49 (1982): 342-49.

held constant, they are relatively more intense in basic research than in applied research or development.

The only independent variable statistically related to the applied research percentage is PROF/SIZE. Its estimated coefficient is positive: apparently, profitability is a determinant of the research portion of a firm's R&D budget rather than the development portion. As seen from column (4), PROF/SIZE is not significantly related to the percentage of R&D allocated to development. The only two variables related statistically to development are FRD/SIZE and CONC.

Federal R&D expenditures were treated as a total amount in the estimation presented in Table 4.5. FRD can also be disaggregated by its character of use. Our survey data allowed for a dichotomy between federal basic and federal applied plus development expenditures. When the equations were reestimated accounting for this division, the results indicated that the composition of federal spending does matter. The positive and significant relationship between FRD/SIZE and both RD/SIZE and D is primarily a result of federal applied plus development expenditures. Alternatively, the FRD/SIZE to BR relationship reflects federal basic allocations.

Equations (4.5) and (4.6) were also estimated using least-squares analysis. The regression results corresponding to equation (4.5) are:

$$\text{PROD} = \underset{(2.19)^{*}}{0.614} - \underset{(-1.31)}{0.427}\,\text{PROF/SIZE} - \underset{(-2.39)^{*}}{0.0013}\,\text{DIV} + \underset{(2.92)^{**}}{0.138}\,\text{FRD/SIZE} + \underset{(0.21)}{0.009}\,\text{MC} - \underset{(-1.83)}{0.002}\,\text{OM} + \underset{(2.21)^{*}}{0.0083}\,\text{CONC};\ R^2 = 0.362. \tag{4.7}$$

**significant at the .01 level
*significant at the .05 level

Of course, the estimated coefficients corresponding to equation (4.6) equal minus one the value of those shown here, and are therefore not reproduced. Profitability is not a significant determinant of the percentage of firms' R&D allocated to product innovation. Similarly, in column (4) of Table 4.5 it was not significantly related to development expenditures. Diversification, however, is a significant variable. Increased diversification decreases the likelihood that a marginal R&D dollar will be allocated to product, as opposed to process, innovation. Presumably, process innovations have the greater probability of multiple usage within the firm. Also, as expected from the results from equation (4.4), the estimated coefficient on FRD/SIZE is positive and significant. Likewise, so is the coefficient on CONC.

THE MARKETING — R&D SYMBIOSIS

In the theoretical discussion and empirical analysis in the two preceding sections of this chapter, we examined firm and industry characteristics associated with the direction of firms' R&D investments. There is another aspect of R&D behavior that needs to be considered, namely, where do ideas originate that later manifest themselves into one form of R&D spending or another?

R&D decisions are not the only decisions made within the organization or the firm. Likewise, the origin of ideas for innovative ventures occurs at many levels, both within the firm and throughout the R&D process. Some innovations are market driven, that is demand induced, while others are technology driven. Those which are market driven may result from direction given to the R&D divisions from the marketing division. The idea that members within a firm complement each other's goals is nothing new to students of organizational behavior, but with the notable exception of Mansfield (1977), little has been written within the economics literature (conceptual or empirical) about the strength of a marketing-R&D symbiosis.

The origins of complementary skills in production can be traced in the economics literature at least to Adam Smith's insights about the benefits to production from division of labor. However, biologist Richard Dawkins notes that this type of mutual cooperation has long existed. He writes, on the basis of his research on insect behavior, that "members of different species often have much to offer each other because they can bring different 'skills' to the partnership" (Dawkins 1976, p. 195). A similar conclusion can obviously be made for the internal workings of the organization.

In our 1982 survey of 146 manufacturing firms (described in Chapter 2), data were gathered on the relative importance of each firm's marketing group as a source of R&D intelligence. Specifically, each firm ranked on a four-point Likert scale, ranging from "always" (= 1) to "never" (= 4), the extent to which their marketing group was used as an informational source in directing R&D activities. The data reflect the role of marketing as an input into the direction of the innovation process rather than as a vehicle for the firm's distribution of its R&D embodied products.

The basic hypothesis investigated here is that the role of marketing is fundamentally more important as a source of R&D intelligence in those firms whose innovative activities are market driven as opposed to technology driven. The difficulty in testing this proposition rests with our inability to quantify the innovation-related competitive environment of the firm. Alternatively, we attempt to infer the nature of this competition from the R&D behavior (an environmental response) of firms. We contend that firms operating in a environment where innovations are demand driven will allocate relatively more of their R&D toward product-related or development activities (than to process-related or basic research) and will pursue an offensive (rather than defensive) posture regarding these R&D ventures.

The empirical analysis presented here is simplistic and primarily intended to illustrate that inter-firm differences in acquiring internal R&D intelligence do exist (and do warrant further study). We correlate each firm's response to the question about the frequency of the use of its marketing department (MARKETING) with the following variables: the percentage of current R&D allocated toward basic research as opposed to development (BASIC); the percentage of current R&D allocated to new or improved products, as opposed to processes (PRODUCT); and the R&D vice-president's perception of the offensive, as opposed to defensive, strategy (STRATEGY) of their firm's product-related R&D (= 1 for "very defensive" and = 5 for "very offensive"). The correlations are reported in Table 4.6.

TABLE 4.6
Correlations between Marketing and the Direction of R&D Activities (t-statistics in parentheses)

	MARKETING
BASIC:	0.158 (2.38)*
PRODUCT:	-0.076 (-1.97)*
STRATEGY:	-0.209 (-2.11)*

*significant at the .05 level

Source: Compiled by the authors.

To the extent that these three R&D-related variables do approximate the market-driven degree of innovative competition facing each firm, the empirical findings are as we predicted. Firms that are more intensively involved in basic research use marketing less frequently (MARKETING = 1 implies the firm always uses its marketing group), and likewise, firms that are more intensively involved in product development use marketing more frequently. Similarly, the more offensive the firm's product R&D strategy (STRATEGY = 1 implies "very defensive") the more it values marketing as a source of R&D intelligence.

Although these summary statistics are far from conclusive, they do suggest that inter-firm differences in the use of marketing as a useful source for R&D intelligence are significantly related to the composition of R&D. Not only is R&D heterogeneous in its nature, but there are heterogeneous forces within the firm that influence the allocation of R&D.

CONCLUSIONS

It is well known that many different activities are classified under the rubric of R&D. However, few researchers have empirically taken this fact into account. The findings presented in this chapter should be interpreted cautiously because of the response subjectivity present in the measurement of the dependent variables. Nevertheless, they represent a beginning attempt to analyze some determinants of the composition of R&D.

One important finding discussed here is the complementarity between federally financed and privately financed R&D. Given that private R&D spending is coming under increased scrutiny as a policy target variable for increasing productivity (as will be discussed in Chapters 6 and 7), apparently one way to stimulate private R&D is by "priming the pump" with federal dollars. However, our results show quite clearly that federal R&D does not lead to increases in all categories of R&D spending, only toward increases at the development end of the continuum.

Care must be taken in generalizing about the causes and consequences of innovative spending until more is known about the microeconomics of R&D. In the following chapter the productivity-related consequences of various types of R&D are examined. There we demonstrate that R&D spending cannot be viewed as homogeneous in either its origin or its character of use if the productivity to R&D relationship is to be fully understood.

NOTES

1. *Business Week* and Compustat are the most comprehensive sets of published data on R&D spending by industrial firms.
2. For a review of these, and related studies, see Kamien and Schwartz (1982).
3. A third interpretation is also possible (Horowitz 1961). Firms with small or falling profits may feel pressure to innovate as a competitive strategy, thus they will increase their innovative outlays.

5

Productivity Growth and Investments in Technology

INTRODUCTION

"Efficient production of goods and services is a primary goal of economic effort, and statistical measurement of productivity is an important tool for monitoring and promoting its advance" (Moss 1979, p. 276). Few economists, researchers, or policy makers would disagree with this evaluation of the concept of productivity; however, some may suggest that this statement underplays the economic and social importance of productivity growth. Productivity is not only a useful barometer of economic conditions, but it is also a fundamental contributor toward our overall well-being. According to Fabricant (1978, p. 502), for example, "higher productivity is essential to our economic growth and [is] a key to a higher standard of living for all our people."

Productivity growth enhances standards of living in several ways. First, our nation's competitive financial and military position within the international community improves with relative increases in production efficiency since more goods and services are produced while scarce resources are conserved. Second, the income generated from productivity improvements can potentially be reallocated toward more leisure time or toward improving conditions of social concern such as environmental pollution and health and safety. And third,

Earlier versions of the materials presented in this chapter were first published in *The Transfer and Utilization of Technical Knowledge* (Link 1982a), *The Bell Journal of Economics* (Link 1982f), *Economics Letters* (Link 1983a) and *Measurement and Analysis of Productivity Growth: A Synthesis of Thought* (Link 1983c).

productivity growth may exert a positive influence in ameliorating inflationary pressures. Rees (1980), among others, argues that when productivity growth slows, economic and political bargaining for income shares results, which exerts upward pressures on prices. These pressures are then intensified as monetary and fiscal policies become the tools of participants in the political arena for sustaining the growth of their own real incomes.

Because of the social importance of productivity growth, the persistent slowdown in productivity, measured as output per unit of labor hours, since the mid-1960s has caused considerable concern. Between 1948 and 1966 the average annual rate of growth in labor productivity in the economy was 3.2 percent; 2.3 percent from 1966 to 1973; and 0.7 percent from 1973 to 1981. Recently, as we came out of a recession, productivity growth has been stronger. Why has the rate of productivity been slowing? There are many factors responsible, in part, for this decline such as changes in energy prices, regulations, capital intensity, and worker's hours and attitudes. An additional influence that has been receiving considerable attention is the lack of growth in R&D expenditures from about 1968 to 1975.

The relationship between productivity growth and R&D spending is investigated in this chapter. First, we present the so-called facts about the productivity slowdown by documenting patterns of measured productivity growth within the U.S. economy. Second, we briefly review arguments commonly associated with other explanations for this slowdown. Third, we present a theoretical framework relating productivity growth to R&D activities, and then finally, we test empirically the predictive powers of this model.

THE PRODUCTIVITY DILEMMA

There are many methods for measuring or indexing productivity. Both partial and total factor, or multi-factor, productivity measures are commonly used. Although the methods employed to calculate these measures are themselves a source of debate, there are some generally accepted productivity statistics that are believed to indicate underlying trends in productivity growth.[1]

There are two usual ways to quantify the efficiency exhibited in the input to output process. Total factor productivity (TFP) refers to the ratio of output to the combination of all inputs, and partial factor productivity refers to the ratio of output to the amount of a single input. If the underlying production process is represented by writing output, Q, as a function of inputs, X:

(5.1) $Q = f(X)$,

then total factor productivity can be represented by:

(5.2) $TFP = Q/\Sigma w_i X_i$,

where the w_i's are input weights ($0 \leqslant w_i \leqslant 1$) and the X_is are the separate input quantities. Partial factor productivity can then be denoted as Q/X_i. For example, labor productivity (Q/L) and capital productivity (Q/K) measures are usually calculated by dividing a measure of real output by either total labor hours or the value of capital inputs, respectively. Total factor productivity (TFP) is similarly derived: the denominator is a weighted sum of both labor and capital input quantities.

The long-term behavior of these three productivity indices for the U.S. private business economy is shown in Table 5.1. Therein, average annual productivity growth rates are reported for selected periods from 1800 to 1973. With the exception of a slight slowdown in the growth of capital productivity between 1948 and 1973, there is no apparent evidence that the economy's overall rate of productivity growth has slackened.

However, when the post-World War II period is considered in more detail, a slowdown in the mid-1960s and 1970s becomes noticeable. Table 5.2 reports average annual productivity growth rates for five sectors of the economy during three separate time periods. For each sector of the economy, except farming, the average growth in productivity (measured by any of the three indices) declined sometime after 1966. In the farming sector there was a significant rise in capital productivity after 1966, but this rise was offset by a decline in labor productivity.

A decline in productivity growth dating from the mid-1960s trend is consistent with the data in Tables 5.3 and 5.4. For each sector reported in Table 5.3, the growth rate in labor productivity increased between the 1948 to 1957 and 1957 to 1968 periods; however, the average annual rate of growth slowed in each of the two subsequent time periods covering 1968 to 1979. A similar pattern is observable from the TFP series reported in Table 5.4. There, growth in the last time period (1973-76) was even reported to be negative. These patterns of declining growth have suggested to some that there may have been *two* productivity slowdowns: one from the mid-1960s to about 1973, and a second thereafter.

Declines in productivity growth become even more pronounced when the statistics are examined at the industry level. For labor

TABLE 5.1

Average Annual Productivity Growth Rates: U.S. Private Business Sector (in percent)

	Alternative Productivity Indices		
Time Periods	*Output/Labor*	*Output/Capital*	*Total Factor Productivity*
1800-1855	0.5	-0.1	0.3
1855-1890	1.1	-0.6	0.3
1899-1919	2.0	0.7	1.7
1919-1948	2.4	1.6	2.2
1948-1973	3.1	1.3	2.4

Source: John W. Kendrick, "Survey of the Factors Contributing to the Decline in U.S. Productivity Growth." In *The Decline in Productivity Growth* (Federal Reserve Bank of Boston, 1980), p. 2.

productivity, the average annual rate of growth in eight of the ten major industry groupings reported in Table 5.5 declined after 1966. The more dramatic slowdowns are seen in the mining, construction, and utilities industries. The exceptions are communications and government enterprises. If total factor productivity measures are examined, a similar pattern can be seen (Kendrick 1980a).

Within the manufacturing sector, by far the largest single sector of the economy, the average annual rate of growth in total factor productivity increased from 1953 to 1966, slowed between 1966 and 1969, and then slowed even more after 1973. As seen in Table 5.6 there was a slight recovery during the 1969 to 1973 period, but the average annual growth rate after 1973 was the lowest ever since World War II. The average annual rate of labor productivity growth declined around 1966 in 19 out of 20 industry groupings, and was at an even lower level after 1973 in 13 industry groupings.

It is clear from these statistics that productivity growth has slowed in the major sectors of the economy since the mid-1960s and surely since the early 1970s.[2] In the following section of this chapter the many causes and correlations associated with the productivity slowdown are reviewed.

TABLE 5.2
Average Annual Productivity Growth Rates: Selected Sectors (in percent)

	Q/L			*Q/K*			*TFP*		
Sectors	*48-76*	*48-66*	*66-76*	*48-76*	*48-66*	*66-76*	*48-76*	*48-66*	*66-76*
U.S. Private Business Sector	3.0	3.5	1.9	1.1	1.5	0.3	2.3	2.9	1.4
Nonfinancial Corporate Business Sector	2.6	3.1	1.8	0.9	1.4	0.0	2.2	2.7	1.3
Farm Sector	5.0	5.3	4.5	-0.2	-0.6	0.6	3.0	3.5	2.2
Nonfarm/non-manufacturing Sector	2.4	3.0	1.4	1.1	1.5	0.5	2.0	2.5	1.1
Manufacturing Sector	2.7	2.9	2.2	0.3	1.0	-0.9	2.1	2.5	1.4

Source: John W. Kendrick and Elliot S. Grossman, *Productivity in the United States: Trends and Cycles* (Baltimore: Johns Hopkins University Press, 1980), p. 34.

TABLE 5.3
Average Annual Labor Productivity Growth Rates: Selected Sectors (in percent)

	1948-79	*1948-57*	*1957-68*	*1968-73*	*1973-79*
Total Private Economy	2.56	3.16	3.11	2.19	0.97
Private Business Sector	2.49	3.02	3.14	2.15	0.75
Nonfarm Business Sector	2.08	2.31	2.77	1.89	0.64
Nonfinancial Corporate Business Sector	2.29	2.61	2.80	2.18	0.98

Source: Martin N. Baily, "Productivity and the Services of Capital and Labor." *Brookings Papers on Economic Activity* 1 (1981), p. 3.

TABLE 5.4
Average Annual Total Factor Productivity Growth Rates
(in percent)

	1948-76	*1948-57*	*1957-68*	*1968-73*	*1973-76*
Total Private Economy	1.64	1.80	2.38	1.21	-0.87
Private Business Sector	1.14	1.57	1.51	0.68	-0.70

Source: Martin N. Baily, "Productivity and the Services of Capital and Labor." *Brookings Papers on Economic Activity* 1 (1981), p. 3.

TABLE 5.5
Average Annual Labor Productivity Growth Rates by Industrial Sector
(in percent)

	1947-66	*1966-77*
Private Business Sector	3.1	1.6
Farm Sector	5.6	4.6
Nonfarm Sector		
Manufacturing	3.1	2.2
Mining	4.3	-1.0
Construction	3.0	-1.5
Transportation	3.1	2.0
Communications	5.2	5.6
Utilities	6.3	2.2
Wholesale and retail trade	2.6	1.9
Finance, insurance and real estate	0.9	0.4
Services	1.5	1.1
Government enterprises	-0.7	-0.1

Source: National Research Council, *Measurement and Interpretation of Productivity* (Wash., D.C.: National Academy of Sciences, 1979), p. 156.

DETERMINANTS OF THE PRODUCTIVITY SLOWDOWN

It has been argued that the slowdowns referred to in Tables 5.1 through 5.6 may simply be statistical artifacts. Few researchers would contend that total factor or partial factor productivity measures are free from either conceptual or data problems. Arguments about the significance of artifacts abound in the literature and are therefore only reviewed here. Even with knowledge of these data problems, partial factor productivity measures, for example, remain widely used. Why? According to Christainsen and Haveman: "Although [these] productivity measures . . . have serious weaknesses, the picture of productivity change which they yield is not greatly different from that of more complete measures" (1980, p. 3).

Cyclical Losses. As a starting point for discussing the determinants of the slowdown, it may be useful to address whether the sources associated with the productivity slowdown are cyclical (that is, short run due to changes in the composition of demand or in the utilization of inputs) or secular in nature (that is, long run due to sectoral or demographic shifts in labor or changes in capital or technological investment behavior). Following others (Perry 1971, 1977; Nordhaus 1972), Gordon (1979) and Mohr (1980) contend that short-run cyclical movements in labor productivity are to be expected. The hypothesis is that firms consistently hire more workers in the last stage of business expansion than is justified by the level of output; thus Q/L falls. As the cycle turns down, labor hiring adjustments are made. Employers can usually make output adjustments at the end of an expansion faster than labor adjustments because of "sticky" contractual agreements. Gordon's (1979) empirical analysis seems to confirm that at least a portion of the post-1965 slowdown in labor productivity is simply due to cyclical "end-of-expansion" phenomenon brought about by managerial misperceptions regarding labor's hiring.

Most of the empirical research conducted, and reviewed below, is based on an implicit maintained hypothesis that the productivity (labor and total factor) slowdown is due to secular changes brought about by either shocks to the economy or by "imposed constraints" on economic activity. Some of these constraints include inflation, high energy prices, and regulations.

Inflation and Energy Prices. There is a high negative correlation between inflation and labor productivity in the economy that can be documented using data as far back as the mid-1940s (Clark 1981, 1982). According to some, this relationship is more than an empirical

TABLE 5.6
Average Annual Total Factor Productivity Growth Rates by Manufacturing Industry (in percent)

	1948-53	*1953-57*	*1957-60*	*1960-66*	*1966-69*	*1969-73*	*1973-76*
Manufacturing	2.9	1.0	1.1	3.9	0.9	2.7	0.1
Food	3.3	2.5	1.1	4.0	1.1	2.8	3.7
Tobacco	1.1	3.5	4.8	2.0	3.6	3.0	1.1
Textiles	0.8	3.6	1.9	8.2	0.1	2.7	0.7
Apparel	2.8	1.4	1.9	2.0	0.8	5.5	2.5
Lumber	0.4	5.8	1.5	7.2	1.6	4.9	-4.7
Furniture	2.2	2.7	0.1	2.7	2.0	1.0	0.8
Paper	3.7	-0.4	1.7	2.8	2.7	5.3	-3.5
Printing, publishing	2.2	2.8	0.6	3.1	0.2	0.7	-1.0
Chemicals	1.8	4.3	2.5	5.0	2.9	4.7	-0.9
Petroleum	1.8	0.6	5.4	4.1	0.8	2.3	-1.7
Rubber	2.1	-2.4	5.7	3.6	3.2	1.4	-1.5
Leather	-2.0	0.7	3.0	3.1	-0.3	2.1	1.2
Stone, clay, glass	2.4	0.1	1.1	2.4	0.8	1.7	-0.9
Primary metals	3.2	-1.5	-4.1	3.3	-3.1	1.8	-3.9
Fabricated metals	1.4	0.3	2.0	2.6	1.5	0.9	-0.9
Machinery, excluding electrical	2.5	-1.9	1.1	2.6	-0.2	2.3	-0.5
Electrical machinery	4.4	2.0	2.6	6.2	2.9	3.7	1.6
Transportation equipment	3.2	1.5	3.3	4.2	-0.5	2.7	3.0
Instruments	4.6	0.6	3.0	3.5	3.1	-0.4	0.1
Miscellaneous	4.0	3.3	2.6	1.6	3.1	2.8	1.6

Source: John W. Kendrick and Elliot S. Grossman, *Productivity in the United States: Trends and Cycles* (Baltimore: Johns Hopkins University Press, 1980), p. 46.

coincidence. Inflation is not a new factor inhibiting productivity: the 1979 report by the Council on Wage and Price Stability, for example, cited inflation as a primary contributor to the post-1973 productivity slowdown. And, there are sound theoretical reasons to expect inflationary tendencies to have a dampening impact on productivity growth.

First, the certainty about price signals is lower during periods of unanticipated, or even prolonged, inflation than in periods of stable prices. Therefore, managerial decisions are made in an above "normal" climate of uncertainty (Jarrett and Selody 1982). This will result in a real efficiency loss because planning horizons are shortened (Hayes and Abernathy 1980). More specifically, the forecasting or actual decision making that takes place in this shortened time frame may be in error. For example, as input prices rise during inflationary periods, it becomes increasingly difficult to determine the portion of the increase in wages, say, that is inflationary or that is actually due to a change in relative factor costs (Clark 1981).

Second, as a result of this increased uncertainty in factor-related decisions, managerial talents may be diverted from long-run cost reducing activities to more short-run production-oriented decisions (Clark 1981). The ramifications of this type of internal organizational shift may show up in an altered attitude on the part of managers regarding their proclivity for risk taking. This factor will not be realized in short-term productivity trends but will surely influence long-run behavior.

Third, not only may inflation indirectly affect the optimal choice of input mix, it may have direct impacts on capital investments. Because the depreciation and replacement of plant and equipment is on the basis of historical costs, prolonged inflationary periods will significantly increase the deviation between these historic costs and effective replacement costs. As a result, current profits, and thus current profit taxes, will be "too high" vis-à-vis that level necessary to permit financing required investments (Lee 1980). The net result is a slowing of the growth of newer vintages of capital to complement labor, and thus, a slowing of the growth of output.

One obvious factor associated with the record of inflation in the 1970s is the 1973 energy crisis. Not by coincidence did some writers quickly contend that the 1973 crisis was the primary factor to bring about the post-1973 productivity slowdown. Siegel (1979, p. 60) in fact is bold enough to state that "energy prices stand out as the single most important contributor to the 1973 [productivity] break."

Why might unexpected or rapid energy price increases alter productivity? A current thesis is that the so-called energy shock represented a structural change in production activity that rendered much existing capital obsolete. On the other hand, the works of Baily (1981a), Coen and Hickman (1980), Hudson and Jorgenson (1978a, 1978b), Rasche and Tatom (1977a, 1977b), and Siegel (1979), using alternative methods, conclude that increased energy prices have significantly reduced measured productivity. Denison (1979), Norsworthy, Harper, and Kunze (1979), and Perry (1977) find only mild negative effects. On the other hand, Berndt (1980b), Clark (1982), and Stein (1981) totally dismiss this energy shock hypothesis. Clark (1982), for example, argues that over time organizational structures (and thus organizational efficiency) have been improving as management learns-by-doing in an atmosphere of low energy prices. The energy price rise may have simply stopped this particular learning process cold, and as a result productivity fell.

Capital Investments. Quite frequently, changes in the growth of capital are viewed as a potentially independent causal factor in explaining the productivity slowdown. When such a hypothesis is tested empirically, most researchers find that the post-1973 slowdown in the capital to labor ratio is significantly correlated with both partial and total factor productivity growth (Baily 1981a, 1981b; Clark 1982; Kendrick 1980c; Klotz, Madoo, and Hansen 1980; Kopcke 1980; Norsworthy and Harper 1981; Norsworthy, Harper and Kunze, 1979; Stokes 1981). Mark's (1978) analysis does not support the findings from these other studies. Economic theory clearly predicts that the marginal product of labor will fall as the absolute stock of capital declines, holding the quality of labor, among other things, constant. Why did the capital to labor ratio fall after 1973? If it were a profit-motivated, factor-usage response to the recently increased price of energy, then treating "energy" and "capital formation" as independent phenomena is incorrect. The interaction of the two effects appears to be a relevant causal factor to consider. If unexpected inflation did create an atmosphere of uncertainty (accompanied by rising interest rates that led businessmen to slow down their long-run financial commitments), then again the interaction of these effects should be considered. To our knowledge, this type of modeling has not occurred. Nevertheless, Fraumeni and Jorgenson (1981, p. 49) argue that the "revival of capital formation is clearly the key to reestablishment of postwar trends in economic growth in the United States."

Regulations. One might also hypothesize that federal regulations, such as environmental and worker safety programs, will reduce measured productivity growth in the regulated industry because the compliance costs are born by diverting real resources (financial, technical, and human) from productive (in a measurable or market sense) activities (Abramowitz 1981; Denison 1979; Myers and Nakamura 1979). It is important to emphasize the adjective "measured" when speaking about the impacts from federal regulations on productivity. Improvements in the value and quality of life resulting from environmental and worker safety programs are simply not quantified in the output numerator of conventional productivity indices. Moreover, the stimulus to supplier industries to provide capital goods and services needed by the regulated industry may increase productivity in the supplier industries. If this occurs, it will appear in a calculated productivity index for the supplier industries as an "unexplained" residual rather than a consequence of regulation.

If the shifted funds come at the expense of newer vintages of plant and equipment then the effective capital-to-labor ratio will be falling absolutely over time and measured productivity growth will likewise fall. If the reallocation comes from other resources, say from total R&D expenditures, then several things could happen to reduce productivity growth. One, the overall level of a firm's indigenous innovative activity will fall; the link between R&D expenditures and total factor productivity is well documented, as will be discussed below. Two, the composition of the firm's R&D portfolio will change, probably more toward defensive R&D (Rothwell 1981) or more toward development (Link 1982f). If either case, one would not expect the newly undertaken R&D to significantly add to the firm's stock of innovation-relating knowledge.

Available data seem to support these propositions. At the aggregate level, the results from Christainsen and Haveman (1981) suggest that between 12 and 21 percent of the slowdown in the growth of measured labor productivity within the manufacturing sector between the periods 1958-65 and 1973-77 is due to federal regulations per se. The study by Evans (1978) supports this view, too. Other specific industry studies also find that regulation, specifically environmental and health and safety policies, have resulted in lower measured labor and total factor productivity growth during the 1970s vis-à-vis earlier periods (Boden, Zimmerman, and Spiegelman 1981; Crandall 1981; Gollop and Roberts 1981; Priest 1981). In fact, Crandall (1981, p. 368) concludes that there "can be little doubt that [labor] productivity has declined more sharply in pollution-control-affected industries than in others."

An alternative to this regulation-induced diversion of resources hypothesis is that the impact of regulations on a firm simply renders certain production methods, and hence certain stocks of capital, cost inefficient. In those instances, no resource reallocation would take place because the compliance costs are prohibitively high. As a result, regulations themselves could lead to a reduced capital-in-use to labor ratio: the end result would of course be lowered measured productivity (although perhaps an increase in environmental quality). Here again it is difficult to separate the impacts of regulation from a declining capital-to-labor ratio during the 1970s. These two presumed causal factors may themselves be related.

Unions. The traditional monopoly view of unionism predicts that unions will decrease labor productivity. Not only will union wage gains in excess of labor productivity increases lead to an inefficient factor mix, but also unions will lower productivity by reducing management flexibility, introducing inefficient work rules, and limiting compensation based on individual productivity. Following this line of reasoning many studies referenced above include a union variable as an independent regressor. Consistently, researchers have reported a negative relationship between levels of industrial unionization and growth in the associated industry's (or individual firm's) total factor productivity (Hirsch and Link, forthcoming).

In contrast to this monopoly concept of unions, a body of literature by Harvard economists has recently emerged that emphasizes a "collective voice/institutional response" (CV/IR) view (Freeman and Medoff 1979b). This view suggests that unionism may increase labor productivity in those work environments characterized by imperfect monitoring, worker complementarities, and attendant public good and externality problems. In this view, unions act as agents for workers by providing a collective voice in bargaining. Productivity is enhanced through decreased turnover and the establishment of grievance procedures, seniority systems, work rules, and the like. In addition, unionization "shocks" management so that existing X-inefficiency is reduced. Thus, the CV/IR view of unionism predicts that unions may increase productivity, even after accounting for the microeconomic adjustments made by firms in response to union wage gains (in particular, an increase in the capital-to-labor ratio and the selection of higher quality workers).

One would be hard pressed to argue that unions have had a significant influence in encouraging the post-1965 slowdowns. It is true that union membership declined throughout that period, but no data exist on the rate of change in the overall effectiveness of unions on firms' production processes.

Entrepreneurship. Although there have been many conceptualizations of who the entrepreneur is and what he does (Hébert and Link 1982), one important, and generally accepted, characteristic of enterpreneurship is an ability to deal with disequilibrium situations (or in a more narrow vein, competitive forces). In a dynamic economy environments change, modes of competition change, and constraints on all activities are constantly in a state of flux.

> Disequilibria are inevitable in [a] dynamic economy. These disequilibria cannot be eliminated by law, by public policy, and surely not by rhetoric. A modern dynamic economy would fall apart were it not for the entrepreneurial actions of a wide array of human agents who reallocate their resources and thereby bring their part of the economy back into equilibrium. (Schultz 1980, p. 443)

It is this constant readjustment to equilibrium, often as a result of innovative insights and actions, that stimulates productivity and ultimately leads to growth. According to some (Klein 1979), the recent slowdown in U.S. productivity growth can be viewed in terms of a decline in businessmen's ability or desire to deal with disequilibria.

Does today's businessman lack the perception and ability to exercise entrepreneurial talent? Have these abilities been declining over the past decade? Hayes and Abernathy suggest that the "rules of the game" may have changed. What could, in our terms, be called "managerial myopia" may simply be the rational response by businessmen to short-run profit incentives.

> We believe that during the past two decades American managers have increasingly relied on principles which prize analytical detachment and methodological elegance over insight, based on experience, into the subtleties and complexities of strategic decisions. As a result, maximum short-term financial returns have become the overriding criteria for many companies. (Hayes and Abernathy 1980, p. 70)

There is some quantifiable evidence that management's focus has become more short run during the recent years. Not only have investments in total R&D been moving away from the basic/long-term/exploratory ends of the spectrums toward the development/short-term/routine ends, as we noted in Chapter 3, but there is an increasing tendency to purchase (directly or through mergers) technologies rather than to innovate. The reason appears to be straightforward: management's reward system is based today on short-run performance criteria guided by short-run quantifiable profits.

As with most of the above mentioned determinants of productivity growth, the direction of entrepreneurial activities is partially related to other factors. The atmosphere created by the unanticipated inflation and the volatile movements of interest rates during the past several years has not been conducive for long-term or risky kinds of investments. Quite naturally, those investments characterized by high payoff ratios have been undertaken faster than others. Unfortunately, those investments that were postponed are likely to be those kinds of investments that are technology embodying, and hence are related to future productivity growth. The post-1965 change in R&D spending is only one example.

THE PRODUCTIVITY GROWTH TO TECHNOLOGY RELATIONSHIP

Technology is considered by many to be one of the most important factors influencing economic activity and economic growth. In the 1950s, the importance of technological change attracted professional attention as a result of the research by Schmookler (1952), Fabricant (1954), Abramowitz (1956), and Solow (1957). Their findings supported the belief that the historical growth in output per man hour in the economy was predominantly due to the application of new production technologies. Perhaps the most influential of these studies was Solow's, which reported that nearly 87 percent of the growth in productivity in the United States between 1909 and 1949 was due to technological change, rather than to growth in capital and labor inputs. There is little doubt as to why technology often became cited as one of the determinants of the post-1965 productivity slowdown. In fact, the motivation for analyzing changes in technology is perhaps justified better within the economic literature than any of the hypothesized factors discussed above.

For nearly two decades economists have been documenting the productivity-to-technology relationship in studies focusing on R&D spending (primarily using cross-sectional industry and firm data within the manufacturing sector). Beginning with Terleckyj (1960), a number of studies were conducted relating productivity growth to, among other things, R&D investments in the stock of technical capital (Minasian 1962; Mansfield 1965; Griliches 1973; Terleckyj 1974). Although the intended purpose of these early works was not policy oriented in the sense of identifying a culprit for the productivity slowdown begun in 1965, the framework used in these studies has become the "state of the art" for many ongoing investigations. It is not surprising then that R&D had come into focus as one primary candidate for explaining the post-1965 and post-1973 slowdowns. A simple analysis of aggregate data reveals that the curtailment in R&D spending is closely associated with the decline in productivity growth: total R&D as a percentage of GNP peaked at 3 percent by 1964, then fell to 2.3 percent by 1975, and even to a slightly lower level by 1977 from which it has subsequently increased back to about 2.5 percent. This ratio reflects sizable cuts in defense and space R&D spending by the federal government between the late 1960s and mid-1970s. Also, industrially-funded R&D (which is the kind of R&D whose link to productivity growth has been repeatedly demonstrated) rose fast in the 1960s in real terms, only in fits and starts between 1969 and 1975, but fairly steadily since.

The available empirical evidence is fairly consistent in its implication that there is a positive direct relationship between total R&D expenditures and measured productivity growth at both the industry and firm level (Link 1980, 1981b, 1981c, 1982a, 1982d, 1982f, 1983a; Griliches 1980a, 1980b; Mansfield 1980; Terleckyj 1980, 1982; Clark and Griliches 1981; Griliches and Lichtenberg 1981; Griliches and Mairesse 1981; Scherer 1982). From these cross-sectional studies it is generally concluded that a similar relationship would exist over time if such data were available. Hence, the slowing of R&D remains a candidate for explaining a portion of the observed productivity slowdown. The only time-related study to verify this belief is by Levy and Terleckyj (1983).

The Analytical Framework

The analytical model underlying these R&D related productivity studies is that the firm (our unit of analysis) operates according to a three-factor production function:

$$Q = A\,F(L, K; T), \tag{5.3}$$

where Q represents output, A is a neutral disembodied shift parameter, L and K are measures of the stock of labor and physical capital, respectively, and T is the stock of technical capital or technical knowledge. T, in turn, is written as a function of research-specific technical capital (C), and other factors (O) that affect its production:

$$T = G(C, O), \tag{5.4}$$

where research capital (C) is some weighted accumulation of previous R&D investments (R):

(5.5) $$R = \Sigma a_i R_{t-i}.$$

These accummulation weights (a_i) reflect, in theory, the influence of both a distributed lag effect of i periods on past R&D and the rate of obsolescence for research capital.

If the underlying production function has the form of a Cobb-Douglas function, then we can write:

(5.6) $$Q = A_0 e^{\lambda t} L^{\beta} K^{(1-\beta)} T^{\alpha},$$

where A_0 is a constant, λ is a disembodied rate-of-growth parameter, and β and α are output elasticities. Constant returns to scale are assumed only with respect to labor (L) and capital (K).

Differentiating equation (5.6) with respect to time (t), it follows that:

(5.7) $$\dot{Q}/Q = \lambda + \beta(\dot{L}/L) + (1 - \beta)(\dot{K}/K) + \alpha(\dot{T}/T),$$

where the dot notation refers to a time derivative, $\dot{Q} = dQ/dt$. Residual productivity growth is defined as:

(5.8) $$\dot{A}/A = \dot{Q}/Q - \beta(\dot{L}/L) - (1 - \beta)(\dot{K}/K).$$

Substituting $\dot{A}/A$ from equation (5.8) into a rearranged version of equation (5.7), equation (5.9) is obtained:

(5.9) $$\dot{A}/A = \lambda + \alpha(\dot{T}/T).$$

The parameter α in equation (5.9) is the output elasticity of technical capital:

(5.10) $$\alpha = (\partial Q/\partial T)(T/Q).$$

Substituting the right-hand side of equation (5.10) into equation (5.9), and rearranging, we have:

(5.11) $$\dot{A}/A = \lambda + \varrho(I_T/Q),$$

where $\varrho = (\partial Q/\partial T)$ is interpreted as the marginal product of technical capital, and $I_T = \dot{T}$ is the net private investment of the firm into the technical capital stock.

If equation (5.11) is stochastic, rather than deterministic, it can be written as:

$$(5.12) \qquad \dot{A}/A = \lambda + \pi (I_T/Q) + \varepsilon,$$

where ε is assumed to obey all classical assumptions. The empirical studies noted above have been based on an estimating equation like (5.12). The underlying starting assumption is that the firm's net investments in the stock of technical knowledge, I_T, can be approximated by the firm's total R&D expenditures. This assumption, probably reflecting data availability, is useful as a starting point, but it is overly restrictive on at least two counts. First, as we emphasized in Chapters 3 and 4, R&D expenditures are not homogeneous in their nature. When I_T is approximated by an R&D flow, one is assuming that all forms of R&D impact on the production process in our identical fashion. This lacks intuitive appeal and, indeed, is theoretically incorrect. For example, the way in which equation (5.12) is formulated implicity assumes that technical knowledge is process augmenting, that is, technical knowledge leads to a shift in the production function. The result of product-related R&D activities will therefore not be captured fully in a index like $\dot{A}/A$.

Second, as we emphasized in Chapter 2, the technological basis of the firm (and hence the industry) is broader than its own R&D program. Technologies can enter the firm not only as the result of indigenous R&D — we have called these induced innovations — but also through such channels as the licensing of other's technologies or the purchasing of new capital equipment that embodies new technologies — we have called these purchased innovations.

The analysis in the following section examines the impact of both inter-firm technology flows and R&D, disaggregated by its character of use, on productivity growth.

The Empirical Analysis

To establish a benchmark for comparison, a stochastic version of equation (5.12) is estimated for I_T approximated simply as a firm's total R&D expenditures. Data from the 1978 survey for 329 manufacturing firms form the basis of the empirical analysis in this chapter. $\dot{A}/A$ in equation (5.12) is measured as the growth rate in residually calculated productivity between 1975 and 1979. First, a productivity index, $[g_t = \ln Q_t - b \ln L_t - (1 - b) \ln K_t]$, was calculated for each firm, t = 1975-1979. The principal data source for these calculations is Compustat. Output, Q, is measured as net sales, defined as gross sales and other operating revenue less discounts, returns, and allowances, deflated by a Bureau of Labor Statistics' industry-specific producer price index. Labor, L, is represented by the total number of employees as reported to stockholders. Capital stock, K, is approximated by the value of gross plant, representing tangible fixed property such as land, buildings, and equipment, deflated by the Bureau of Economic Analysis' implicit price index for nonresidential gross private investments. The average share of labor in total sales, b, over the period 1975 to 1979 is estimated as the total labor expenditures of the firm in 1977 per unit of 1977 sales. For those firms not reporting labor-related expenditures to Compustat, labor's share is computed using the product of the average 1977 annual wage, from the Bureau of the Census, in each firm's industry and the total number of 1977 employees in the firm. The average share of capital is $(1 - b)$. Then, $\dot{A}/A$ is measured as the slope coefficient from a regression of g_t on trend for each firm.

I_T is first measured as each firm's 1977 total R&D expenditures, RD. Also held constant in the estimation is an index of industry unionization. This index is the percentage of workers unionized in the three-digit SIC industry in which the firm performs its main operations, as reported in Freeman and Medoff (1979a). The least-square results associated with the technical capital variable measured as RD/Q are in column (1) of Table 5.7. The estimated coefficient is positive, as many researchers have found, but here it is small and not significantly different from zero.[3]

Data related to each firm's total expenditures on technological advances are not published; however, a rough indicator of these total activities, TECHADV, can be constructed for each firm using data on the percentage of the firm's technological activities induced through its own R&D activities, INDUCE. (These percentages were discussed and analyzed in Chapter 2.) If INDUCE = RD/TECHADV, then TECHADV = RD/INDUCE. Given data on R&D expenditures, firm specific indicators of TECHADV are calculated for each firm. Then the relationship between productivity growth and these indicators of total expenditures on technological advancements is examined in a second version of equation (5.12): the least-squares results are reported in column (2) of Table 5.7. The size of the regression coefficient on this measure of the I_T variable has increased by almost a factor of two; but more importantly, it is statistically significant.

The third version, whose results are reported in column (3) of Table 5.7, dichotomizes the firm's total expenditures on technological advancements, TECHADV, into those induced, RD, and those purchased, PURCH(= TECHADV − RD). It is not surprising, given the results reported in column (2), that the coefficient on purchased technologies is numerically larger and more significant than the coefficient on indigenous R&D related activities. In fact, this finding complements the results from Terleckyj (1974, 1980, 1982) and Scherer (1982) that were based on aggregate industry data. We view the specification underlying these results as the more complete specification in that inter-firm flows of technology are explicitly held constant.

Still, indigenous R&D is not correctly represented. The heterogeneity of R&D related technologies is further conceptualized in two ways. First, indigenous R&D expenditures are dichotomized into those going toward basic research, BR, and those going toward applied research and development, AR&D. Second, total R&D expenditures are dichotomized between product- and process-related activities, PROD and PROC. Data for these four categories of spending come from inputing the percentages for each category (which were analyzed in Chapter 4) to total 1977 R&D expenditures.

The least-squares results presented in column (4) of Table 5.7 correspond to the basic research/applied research plus development division of indigenous R&D. The estimated coefficient on the basic research variable is highly significant and quite large in size. In contrast, the statistical influence of applied research plus development expenditures conducted within the firm have a lesser influence on the firm's productivity growth. Similar findings are reported in column (5) for product- and process-related expenditures. R&D directed to process-augmenting technologies are more highly correlated with productivity growth than are those R&D expenditures allocated toward product modifications, holding purchased technology flows constant. Not only is the regression coefficient much larger on the PROC term, it is also more significant.

It appears from the results reported in Table 5.7 that both induced and purchased technologies are important determinants of productivity growth.[4] Also, induced technologies emanating from nondevelopment kinds of R&D appear to be relatively more important contributors to measured productivity growth. Perhaps the most important generalization possible from this analysis is that R&D activities do influence productivity, but the subtleties of the R&D to productivity relationship begin to be realized when the sources of R&D technologies and the character of R&D activities are considered separately.

The R&D variables considered in the estimations reported in Table 5.7 refer to endogenous, or self-financed, investments. The influence of federally financed R&D on measured productivity growth was also considered. A FRD/Q term was included in each of the estimating equations. Although the regression coefficients are always positive, they are in no instance significant at a level greater than .30

TABLE 5.7
Regression Results from Alternative Specifications of Equation (5.12)
(t-statistics in parentheses)

Independent Variables	*(1)*	*(2)*	*(3)*	*(4)*	*(5)*
RD/Q	0.082 (1.48)		0.066 (2.23)*		
TECHADV/Q		0.118 (2.40)*			
PURCH/Q			0.244 (3.11)**	0.211 (2.93)**	0.277 (3.01)**
BR/Q				2.09 (3.92)**	
(AR&D)/Q				0.016 (2.01)*	
PROD/Q					0.036 (1.89)
PROC/Q					0.522 (2.68)**
R^2	0.33	0.41	0.49	0.58	0.52

**Significant at the .01 level.
*Significant at the .05 level.
Source: Compiled by the authors.

(results are not shown). Even when a subsample of the 329 firms containing only those firms receiving federal R&D, similar results were found. When federal R&D expenditures are dichotimized into the two categories, basic research and applied research plus development, the estimated coefficient on the federal basic term becomes significant at the .10 level. This finding should be interpreted as tentative (Link 1981b). Link (1982d) argues that federal R&D not only complements private R&D spending (as shown in Chapter 4) but also impacts on the productivity of private R&D. In other words, the estimated rate of return to private R&D, reported in Table 5.7, may not be independent of the amount of federal R&D received by the firm. Link shows that firms receiving federal R&D have a lower marginal return on their own R&D. Perhaps slower productivity growth is an opportunity cost associated with federal contracting.

As was discussed in the section of this chapter on "Determinants of the Productivity Slowdown," environmental regulations have had an impact on the slowing of productivity growth since the mid-1960s.

One reason for this, as discussed in Chapter 3, is that firms are constrained to allocate scarce resources to comply with these regulations. Presumably, some of the resources allocated toward R&D to comply with regulations lead to activities whose output is not measured by standard productivity indices.

We test this hypothesis using survey data for 97 firms in the chemicals (32 firms), machinery (51 firms), and petroleum (14 firms) industries (as were summarized in Table 3.6). Equation (5.12) was estimated with the R&D variable dichotomized into the portion of R&D allocated toward activities to comply with environmentally related government regulations, RD_E, and to maintain the firm's "traditional" types of innovative activities, RD_T. A unionization variable, U, is again included.

The least-squares regression results, with t-statistics in parentheses, are reported below in equations (5.13) and (5.14):

$$\dot{A}/A = \underset{(2.51)^*}{0.067} + \underset{(1.67)}{0.312}\,(RD/Q) - \underset{(-1.98)^*}{0.103}\,U;\; R^2 = 0.34 \qquad (5.13)$$

and,

$$\dot{A}/A = \underset{(3.01)^*}{0.062} - \underset{(-1.72)}{1.22}\,(RD_E/Q) + \underset{(2.87)^*}{0.538}\,(RD_T/Q) - \underset{(-1.90)}{0.092}\,U; \qquad (5.14)$$

$$R^2 = 0.39.$$

*significant at .05 level

These results suggest that in this smaller sample of firms the relationship between total R&D expenditures (RD/Q) and measured productivity growth is positive, but weak. However, when R&D is dichotomized, the estimated coefficient on RD_T is positive and highly significant while the estimated coefficient on RD_E is negative and marginally significant.[5]

The impact of environmental regulations on the direction of R&D and the opportunities for R&D related breakthroughs might be intrinsically stronger in certain industries than in others. This possibility was also considered by including dummy variables, interacted with the R&D variables, in these two regression models. The results are summarized in Table 5.8. There appear to be marked inter-industry differences in the magnitude of the relationship between R&D and measured productivity growth.[6] The estimated coefficient on (RD/Q) and (RD_T/Q) is largest in the chemicals industry and smallest in the petroleum industry. The estimated coefficient on (RD_E/Q) is smallest in the chemicals industry and largest in the petroleum industry. This pattern is consistent with an additional survey finding that the percentage of environmentally related compliance costs that are covered by R&D activities is, on average, 10.8 percent in the chemicals, 17.9 percent in the machinery, and 23.1 percent in the petroleum industries.

TABLE 5.8
Calculated R&D Coefficients by Industry
(calculated t-statistics in parentheses)

	Independent Variables		
	(1)	*(2)*	*(3)*
Industry	*RD/Q*	*RD_E/Q*	*RD_T/Q*
Chemicals:	0.512	-0.583	0.912
	(2.16)*	(-1.90)	(3.65)**
Machinery:	0.337	-1.01	0.625
	(1.19)	(-1.27)	(2.13)*
Petroleum:	0.205	-2.36	0.241
	(1.31)	(-1.73)	(1.89)

*Significant at the .05 level.
**Significant at the .01 level.
Source: Compiled by the authors.

CONCLUSIONS

The productivity to technology relationship is complex. Technologies enter the firms from various R&D sources and impact differently, both in the level of magnitude and presumably in the lag structure, on productivity growth. But still, wherever these technologies originate their basis is tied to R&D activity.

In this chapter we have attempted to illustrate empirically that factors correlated with the recent slowdown in productivity growth have been not only the slowdown in the rate of growth of industrial R&D spending, but also the change in the composition of R&D (as first documented in Chapter 3) away from both the basic and process-related ends of the R&D continua to the more short-term and applied ends, and, in some cases, toward activities undertaken solely to comply with government regulations.

We therefore conclude that to encourage productivity growth, industrial R&D is a reasonable target variable to consider, as many have done and are doing. However, as we will argue, extreme care must be taken in formulating policy to ensure that incentives given to firms increase those kinds of R&D-related innovative activities most likely to enhance the productivity growth of the economy. Some existing public policy proposals are analyzed in the following chapter.

NOTES

1. For a critical review of the various methods used to calculate productivity indices, see Link (1983c).
2. Nordhaus (1980) points out that if one examines aggregate labor productivity data from 1912 to 1979, the type of post-1973 slowdown the economy is now experiencing occurred twice during the last 60 years. Such slowdowns are simply not uncommon phenomena (and, in his opinion, perhaps not worthy of extraordinary policy debate).
3. Some have interpreted this weak relationship using post-1973 data as an indication that there has been a collapse in the productivity of R&D compared to the 1960s. It is our belief that R&D is as productive as it was in the 1960s, at least at the margin. Rather, the composition of R&D has been changing and hence the mismeasurement biases associated with total R&D expenditures proxying investments in the stock of technical capital have been compounding.
4. In all of the estimations presented in Table 5.7 an index of industry unionization was held constant in the regression equation. The estimated coefficient on this variable was always negative and generally significant at only the .10 level.
5. When RD_T is dichotomized into basic research and applied research plus development expenditures, the estimated coefficient on the basic research term is 2.61 with a t-statistic of 3.96, and the estimated coefficient on the applied research plus development term is 0.317 with a t-statistic of 2.51. Also, the sign on the RD_E term remains negative.
6. A series of F-tests was conducted to determine if the structure of the estimated equations differs between industries. The null hypothesis of equality of coefficients across industries was rejected at the .01 level ($F_{4,88}$ = 7.27) in favor of the specifications reported in columns (2) and (3) of Table 5.8. Also, pair-wise F-tests were conducted for the three industry groups. The null hypothesis of equality of coefficients between the chemicals and machinery industries was rejected at the .01 level ($F_{2,80}$ = 21.79); between the chemicals and petroleum industries the null hypothesis was rejected at the .01 level ($F_{2,40}$ = 10.61); but between the machinery and petroleum industries the null hypothesis was accepted ($F_{2,59}$ = 2.71).

6

Tax Incentives for R&D: A Critical Evaluation

INTRODUCTION

The recent decline in productivity growth in the United States has generated considerable attention, and as a result a number of public policy proposals have been suggested to reverse the declining trend. As we illustrated in the previous chapter, the magnitude of the productivity decline is well documented; however, there is much dispute about the merits of disparate public policy options for reversing this trend. This dispute does not come from the fact that policy makers are naive concerning the many factors contributing to the productivity slowdown, but rather from the general lack of understanding, both in policy and academic circles, regarding barriers to productivity.

Our focus in this chapter is on policy proposals centering on additional R&D activity. Suggested approaches include increasing governmental support for extramural research, setting up new cooperative research centers, and stimulating industrial research through various tax mechanisms. We will concentrate here on only the latter. The study of public policy has until recently been directed to such "policy tools" as grants and only limited attention has been given to subsidies, tax credits, and other tools that entail quasi-market mechanisms or nontraditional public management. Recently there has been a shift away from traditional grants approaches: this shift is nowhere in greater evidence than in R&D policy, a policy domain traditionally dominated by grants and focused on projects. As the Reagan Administration has begun to display its

This chapter draws, in part, from Bozeman and Link (1982).

commitment to cutting domestic spending and revamping policy making, the focus has shifted away from grants assistance (except in basic research) to policies aimed at stimulating private sector R&D. Since programs for government support of R&D are almost all within that portion of the federal budget that is "controllable," the prospects for increased government R&D spending (the portion of total R&D that declined most within the United States during the late 1960s and early 1970s) are uncertain; tax incentives could provide a reasonable substitute.

In this chapter we first review various tax policies designed to stimulate R&D and then we assess the viability of tax incentives as a policy tool. Not only are we interested in the economic rationales that underlie policies and policy tools, we are also concerned with the matters of political and managerial feasibility.

TAX POLICIES FOR R&D

Although many provisions of the U.S. tax code impinge on R&D and innovation, the impact of the tax law is often indirect, highly uncertain, and in some cases virtually unfathomable. For example, some observers, such as Boskin (1981), contend that the constantly evolving personal income tax laws have significant implications for innovation, especially for the entrepreneurial phase of the innovation process. While personal income tax provisions are arguably important for innovative activity, a comprehensive assessment of personal income tax policy is beyond the scope of this chapter. Our focus is specifically on those portions of the tax code that are intended, by design, to influence innovation and/or R&D spending.[1]

While interest in tax incentives for R&D is now high, one should not infer that the idea of using tax policy as a targeted stimulus for R&D activity is new. Tax incentives for R&D have long been part of the U.S. tax code. For several decades large firms have been deducting most of their expenses for ongoing research and experimental (R&E) expenditures as they were incurred. The law was ambiguous in regard to the deductibility of the lesser (but still regular) R&D expenditures of small firms: even if these expenditures were capitalized, the period of amortization was unclear (Slitor 1966).

In 1954 an important step was taken with the adoption of Section 174 of the Internal Revenue Code, which codified and expanded tax laws pertaining to firms' R&D expenditures. This provision permitted businesses to deduct fully research and experimental (*not* development or research application) expenditures in the year they were incurred.

If the firm chooses not to expense research and experimental expenditures in the year incurred, there is an option to capitalize R&E expenditures and amortize them over a period of not less than five years, beginning with the month in which benefits are first realized from the expenditures. Since there is no precise way to determine when benefits are first realized, it is assumed, in the absence of contrary evidence presented by the firm, that benefits are first realized in the month in which the product or process gained from the R&E activity first produces income. The deferral option was designed with the intent of benefiting new small firms that have little or no taxable income during their early years.

The Internal Revenue Service made clear that the expensing option was limited to direct R&E expenditures and was not applicable to expenditures on capital assets necessary to conducting R&E activities. However, the costs of such depreciable assets as machinery, equipment, and facilities are partly recovered through depreciation allowances that apply for investments on any depreciable property.

Expenditures for R&E are given preferential treatment under Section 174, but initially there was little if any indication as to precisely what activities qualified as research and experimentation. Stipulations and limits were provided in subsequent Department of Treasury regulations (especially Treasury Regulation 1.174-2). These regulations usually refer to "research and development costs in the experimental or laboratory sense." These terms generally include "all such costs incident to the development of an experimental or pilot model, a plant process, a product, a formula, an invention, or similar property, and improvements to already existing property" of the type mentioned above. It also includes the cost of obtaining a patent; such as attorney fees incurred in making or perfecting a patent application. Specifically excluded are expenditures for testing or inspection of materials or products for quality control, management studies, advertising, and promotion. A number of tax court decisions have further codified the legal meaning of research and experimental expenditures.[2]

While the immediate expensing of R&E expenditures is generally viewed as a tax incentive, Kaplan (1975) argues that Section 174 is more for administrative convenience than for a tax incentive. While there is a consensus about the salutary effects of R&E expenditures on both innovation and the creation of marketable products, the *marginal* contribution of R&E is difficult to determine as it is closely intertwined with a host of other inputs such as management structure, quality control, marketing strategy, and level of employee creativity:

> Rather than get immersed in estimating the benefits from this whole range of intangible assets, it seems prudent for Congress and the IRS to allow full deductibility of expenditures on such intangibles as an administrative convenience rather than as a tax incentive. If Congress had truly wanted to supply a tax incentive for R&D, it could have allowed an immediate write-off for investment in tangible R&D assets (building, machinery, and equipment). (Kaplan 1975, p. 7)

Indeed, one of the continuing interests of business leaders has been the expansion of tax incentives to include write-offs or tax credits for tangible R&D assets; such a proposal was given the highest priority by the Committee for Economic Development (1980).

While Section 174 is perhaps the most encompassing and widely used tax incentive for R&D, a number of additional elements of the tax code are also viewed as R&D incentives. Individuals and corporations are allowed to deduct contributions to educational and scientific organizations held to be operating in the public interest. There are limits described in Section 170(a) on such deductions: individuals may deduct not more than 50 percent of adjusted gross income for such contributions and corporate deductions are limited to 5 percent of taxable income. Furthermore, the income of scientific and educational organizations operating in the public interest is exempt from federal income tax. According to Section 501(a-c) of the tax code, organizations qualify for this exemption if they are conducting scientific research that is "directed toward benefiting the public." The operating standard is that the work must result in information "published in a treatise, thesis, trade publication, or in any other form that is available to the interested public." If this standard is met, the organization qualifies for the exemption even if the research is performed under "a contract or agreement under which the sponsor or sponsors of the research have the right to obtain ownership or control of any patents, copyrights, processes, or formulae resulting from such research." This provision has attracted much attention, partly in response to the substantial increase in proposals for developing industry-university cooperative research programs.

Another provision viewed as a tax incentive pertains only to individual inventors. An individual inventor may treat proceeds from the sale of a patent as a capital gain instead of ordinary income. To obtain this benefit, the inventor must transfer "all substantial rights": that is, he must sell the patent rights or grant an exclusive license (since a nonexclusive license does not qualify). In such cases the more favorable capital gain tax rate applies, but only for the inventor and not his employer or the purchasers of the rights.

CURRENT PUBLIC POLICIES AND PROPOSALS

As interest in stimulating R&D has increased, a number of policy options have been considered. In this section we review some of these more prominent proposals as well as the relevant provisions of the Economic Recovery Act of 1981. In Chapter 7 we present a rationale for an alternative policy – a tax credit for joint venture R&D.

Legislative Initiatives

One of the more sweeping proposals for stimulating R&D is the Stevenson-Wydler Act. The policies embodied in this act are traditional in the sense that they depend on government management and support of R&D but are nontraditional in the sense that they are not oriented toward grants assistance. The act has a number of objectives and includes a variety of initiatives. Among the more important provisions are: (1) the establishment of an Office of Industrial Technology in the Department of Commerce; (2) the creation of a grants program for Centers for Industrial Technology (so-called generic technology centers); and (3) a requirement that each federal agency with an R&D laboratory commit 0.5 percent of its R&D budget to technology transfer. The current political climate does not seem auspicious for a program that expands government management of R&D and early signs seem to indicate that the Stevenson-Wydler Act will have little impact. Current budget proposals have not included funding for the more important provisions of the act and, moreover, the Congress and the Reagan Administration joined forces in rescinding a good portion of the funds.

The policy proposal that seems more aligned with current administrative thinking is a tax credit for industry (or industry-sponsored) R&D. Congressman Charles Vanik introduced legislation during the 96th Congress that would have provided a 25 percent tax credit to firms for contributions to university research. While this bill generated some support it was not enacted. Moreover, the Vanik proposal, and similar legislation recently introduced by Senator Bradley and Representative Shannon, were passed over in favor of more general policies included in the Economic Recovery Act of 1981.

Professional and Trade Association Proposals

Among the several proposals (Cordes 1981) for stimulating R&D and technological innovation, two stand out. The Committee for Economic Development (CED 1980) monograph, *Stimulating Technological Progress*, and the National Academy of Engineering (NAE 1980) study,

The Impact of Tax and Financial Regulatory Policies on Industrial Innovation, merit special attention because the respective policy statements are coherent and well thought out, and because they take a broad view of the role of tax incentives for enhancing R&D. Each report is similar in its argument that tax incentives must be considered in conjunction, not in isolation, with more general macroeconomic policies, policies for stimulating business investment, and government regulation of industry. At the same time, there is some divergence between the two proposals. For example, the NAE report is much less sanguine than the CED report in regard to the potential of tax incentives for R&D as a stimulus to industrial productivity.

The NAE report assigns a relatively low priority to R&D tax incentives because of the complexity of the innovation process, limited knowledge about the effects of R&D on innovation, and the differential importance of R&D among firms (as discussed in Chapter 2). In some industries (see Table 2.1), R&D expenditures seem to comprise only a modest proportion of the total cost of innovation. While noting that tax incentives for R&D may play a part in a more comprehensive approach to stimulating innovations, the NAE report does not go as far as to offer a specific proposal.

The major tax policy recommendations of the NAE report include the following:

- An expansion of tax incentives for capital spending.
- Further research to assess the impact of Treasury Regulation 1.861 to determine the amount of tax raised by this provision and possible effects suppressing innovation.[3]
- Investigation of the effects of the corporate income tax on small firms and especially new firms. The report suggests that depreciation deductions and investment tax credits have little value to new firms that have not yet been able to generate profits. Carry forward provisions merit further scrutiny.
- Lowering the maximum capital gains taxation for individual investors.
- Increased deductibility of capital losses.
- Tax-free rollover of equity investments in small technological firms.

The CED report takes a strong stand on tax incentives for R&D, expressing the view that there is a strong correlation between industrial productivity and R&D expenditures.[4] While arguing that productivity and innovations are influenced by a great many factors, the CED nonetheless supports a program of selective, targeted tax incentives for R&D as a central thrust of its policy proposals. The report notes that the United States has for some time lagged behind other Western industrial nations such as Japan, West Germany, and France in providing more liberal tax incentives for industrial R&D.

The CED report also proposes several general tax incentives including reduction in personal income tax rates, reduction in capital gains tax, and reduction in maximum marginal tax rate. Unlike the NAE report, the CED proposals include several recommendations for targeted tax incentives including the following:

- A permanent increase, from 10 to 20 percent, in credit for investment in R&D facilities.
- Externally acquired patents and intangible technology are to be subject to accelerated depreciation over a ten-year period or their demonstrated life, whichever is shorter. The reasoning here is that the effective life of a patent often terminates long before the end of its traditional legal life of 17 years.
- Since Treasury Regulation 1.861-8 appears to have adverse effects on innovation, it should be revised such that only directly related R&D expenses need be deducted from foreign source income.
- Corporations should be permitted to deduct a specific portion of their contribution to nonproprietary research conducted at universities.
- Special tax incentives should be developed to encourage investment in pioneer plants. The reasoning here is that the high costs of pioneering (first of-a-kind commercial size facilities) pose special capital formation problems. Tax incentives or cash grants for pioneer plants have been employed in some nations as a means to abet the commercialization of new technology.
- A number of policies pertaining to the investment activity of small business merit attention: increased deductibility of capital losses; extension of carry-back provision from three to seven years; increase in the number of investors (from 15 to 100) allowed to meet the qualifications under Subchapter S of the tax code for small corporations.

Despite the attention given to tax policy, these recommendations are only part of a larger package suggested by the CED. Considerable attention is given to reducing government regulation of industry, reforming patent policy, and even direct federal support for industrial R&D.

The Economic Recovery Act of 1981

One of the most significant, and in some respects controversial, changes in the tax treatment of R&D expenditures and R&D assets was added by Congress to the Reagan Administration's policy centerpiece, the Economic Recovery Act of 1981 (hereafter referred to as the Tax Act). The major changes affecting R&D embodied in the Tax Act include: (1) a faster depreciation of R&D assets; (2) a two-year suspension of Treasury Regulation 1.861-8 to study its impact; (3) an increase in the deduction permitted for contributions of newly-manufactured research equipment to universities; and (4) a tax credit for R&D expenditures.

Since our primary interest in this chapter is in incentives provided by tax credits for R&D, we limit our focus only to the tax credit provisions of the Tax Act. Roughly, the tax credit in question is for R&D expenditures in *excess* of the average amount spent over the last three taxable years. Expenditures qualifying include in-house expenditures for R&D wages and supplies; 65 percent of the amount paid for contract research; 65 percent of corporate grants to universities and scientific research organizations for basic research. Expenses must be paid for by the taxpayer during the taxable year and must pertain to the carrying on of trade or business. The credit is effective for research expenditures paid or incurred after June 30, 1981 and before January 1, 1986 (when in the absence of further legislation the credit will no longer exist. The administration has recently supported a three-year extension of the credit).

The amount of the credit is 25 percent of the excess of the taxpayer's eligible research expenditures for the taxable year, over the average of the taxpayer's yearly eligible research expenditure during the base period. The base period is, in general, three taxable years immediately preceding the taxable year for which the credit is claimed. The base period research expenditures cannot, however, be less than half of the expenditures for the current year. A number of special base period computations have been specified for new firms or firms just beginning to engage in R&D activity.[5]

Several limitations of the Tax Act are worth noting. In the first place, the requirement for "carrying on a trade or business" means that expenses incurred in connection with trade or business but not pertaining to the development of potentially marketable goods and services fail to qualify. In the second place, and perhaps just as important, the credit does not apply to research expenditures paid or incurred prior to commencing a trade or business. Only wages paid for actual research qualify for the credit. Therefore, wages for laboratory scientists and engineers and their immediate supervisors would qualify, but wages for general administrative personnel or other auxillary personnel (such as computer technicians working in a multi-purpose computer and information processing department) would not.

The effects of the tax credit were much debated before adoption and will continue to be a subject of controversy. In fact, a recent McGraw-Hill survey (1982) shows that firms intend to raise R&D spending by 17 percent in 1983 despite the current economic recession. Furthermore, we have reported in Chapter 3 that a panel of R&D vice-presidents in manufacturing firms forecast that the percentage of R&D allocated to basic research or long-term exploratory research will steadily increase through 1990. But, as Press (1982) notes, and as we documented in Chapter 3, the pattern of sacrificing long-term R&D spending to protect short-term investments appears to have been changing well before the Tax Act was formulated.

The Tax Act has not preempted other proposals for R&D incentives and, moreover, there is considerable legislative activity aimed at refining and extending the tax incentives provided for in the act. The R&D Tax Planning Act (S.2256) was recently introduced and would make permanent the 25 percent tax credit as well as the suspension of Treasury Regulation 1.861-8. Additionally, H.R. 4667 was introduced to broaden the definition of institutions eligible under the Tax Act to receive credited R&D donations from corporations. Another proposal, H.R. 5344, would extend benefits to new businesses.

Other Recent Proposals

Two proposals recently advanced, but not yet appearing on a legislative agenda, include those by Klein (1981) and Hufbauer (1981). Both are aimed at rewarding R&D success rather than rewarding actual R&D expenditures or increments to R&D budgets.

The Klein proposal seeks to provide benefits for risk taking while at the same time suppressing inflation, tying rewards to performance and reducing the sensitivity of R&D investment to fluctuation in the business cycle. He summarizes his proposal as follows:

> Assume that one wants a tax that both stimulates productivity advances and constrains wage increases. A tax based upon this logic would reward firms on the basis of improvement of their own price performance relative to the industry's average performance during the preceding year. Suppose that, for the purpose of judging the industry's average performance during the previous year, the firm utilizes price data published by the government. If the firm increases

> its prices at the same or greater rate than the industry as a whole, it will pay taxes at a rate equivalent to the previous year. But, assume that to the extent it improves its price performance, then it will be able to reduce its tax burden considerably. (Klein 1981, p. 174)

The basic rationale of the proposal is that firms should be encouraged to take risks in order to escape higher taxes. It is expected that firms would take greater risks in R&D in order to enhance price performance. Alternatively, lower prices might be made possible by a combination of R&D risk taking and wage constraint.

While the objectives of the Hufbauer proposal are quite similar to Klein's, the approach is different and is more limited in scope. His proposal entails alterations of tax policy pertaining to patents:

> All receipts from any *publicly registered* licensing or sale of patents or know-how to *unrelated* parties should be tax-free. In other words, royalty payments should be tax-free to the seller, but remain deductible to the buyer. This provision would reward R&D success, not merely expenditure. And it would greatly broaden the market for technology. (Hufbauer 1981, p. 148)

We advance an alternative proposal in Chapter 7. Basically, we argue that current policies, including those by Klein and Hufbauer, do not deal adequately with the public goods characteristic of R&D-generated knowledge. We argue that current policies favor firms continuing to invest in R&D nearer to the applied research and development end of the continuum. We propose a tax credit for joint venture R&D: one result of joint venture activities would be to provide an incentive within firms to increase the portion of their R&D allocated toward basic research – the portion of R&D so far most highly correlated empirically with productivity growth (as shown in Chapter 5).

TAX INCENTIVES FOR R&D: EVALUATION OF THE POLICY TOOL

Economists and policy makers have shown little reluctance to jump into the fray; not only have several tax policy proposals been offered but a number of studies have provided assessments of proposals. In this section we are less interested in evaluating any particular proposal than we are in considering the broad advantages and disadvantages of tax incentives, especially tax credits, as a policy tool for stimulating R&D. Naturally, questions pertaining to policy tools cannot be completely divorced from the specifics of policies. However, our primary objective is to provide a broad overview of R&D tax incentives as a device.

There are three fundamental questions that need to be considered when analyzing any policy tool: questions of need, questions of tool selection, and questions of policy design. Is there a need for a policy tool to stimulate R&D (need)? Are tax incentives the appropriate tool for stimulating R&D (tool selection)? Can we design tax incentives that will achieve the desired economic results and at the same time meet criteria of administrative efficiency and political feasibility (policy design)?

Questions of Need

Of primary concern is whether there is a need for the kind of tax credit for R&D provided in the Economic Recovery Act of 1981, or for similar incentives such as those advanced by the NAE, the CED, the Commerce Technical Advisory Board, or the Domestic Policy Review Board of the Department of Commerce (Cordes 1981). However, before this concern is investigated two other questions warrant attention. First, are there general rationales for government interventions in support of innovation and R&D? Second, is there evidence that any *additional* intervention (whether in the form of tax incentives or direct assistance) is required at this time?

The first question has been pondered for some time, and some satisfying theoretical conclusions have been reached. The earliest and most interesting theoretical arguments were presented by Arrow (1962) and Demsetz (1969), among others, but a more succinct summary has been provided by Slitor (1966). Slitor's arguments for government intervention can be summarized as follows:

- Technical advance accounts for a large fraction of productivity growth.
- The private rate of return on R&D is high, but the social return is believed to be even higher.
- The gap between social and private return is largely a function of limited appropriability of results, externalities, limited rights of exclusion; yet property rights should not be greatly extended because the social costs of additional use of technology are negligible.
- Risk and uncertainty inherent in R&D inhibit firms' investments.
- Economies of scale and the need for pooling risk discourage research in smaller firms; nonintervention may result in R&D monopolies for the largest firms.
- There are "gray areas" between public and private sectors that offer limited appeal to the profit motive.[6]

Although Slitor's list summarizes the most widely acknowledged rationales for government intervention, Hufbauer (1981) raises an interesting point concerning the underinvestment in R&D. Granted that industries recognize R&D investments as an activity that involves long-run payoffs, why is there not more of a propensity to look beyond the immediate future? Of course there is much speculation about the effects of rapid executive succession, turnover, and absentee ownership on firms' time perspectives (as we discussed in Chapter 5), but Hufbauer offers an alternative argument:

> A vicious (or virtuous) circle is at work. When profits decline, R&D is disproportionately cut; when profits rise, R&D disproportionately rises. Why? because R&D is expensed annually on the profit and loss statements even though it has the economic character of a long-term investment. (1981, p. 145)

Thus, Hufbauer's conclusion is that current accounting conventions are one cause of firms' tendency to shorten their investment horizons during periods of decline, thereby smoothing earnings per share. There is some plausibility to this explanation but one of its premises is controversial. As we pointed out in Chapter 4, the research on the relationship between aspects of R&D investments and profitability is not conclusive.

The more pressing question is whether there is a need for additional government intervention. From one standpoint this problem has no resolution since there is no available calculus to provide a precise answer to the question, "How much is enough?" Economic theory clearly states that additional government intervention is warranted as long as the marginal benefits from such expenditures are greater than the marginal costs. However, the practical applicability of this criterion in an environment of market failures is questionable. Still, the fundamental question of need cannot be avoided. Unfortunately, the question of need cannot be answered simply by inspecting available R&D statistics. The trends in R&D spending discussed in Chapter 3 provide perspectives for some meaningful dialogue regarding past activities and future goals, but they do not lead inexorably to the conclusion that R&D must be accelerated or that further government intervention – either by direct assistance or by tax incentives – is required.

It seems important to note that the government rationale for encouraging additional R&D expenditures rests on a series of problematic (though perhaps correct, as partially evidenced from our analysis in Chapter 5) assumptions. The logic can be broken into the following three assumptions:

- Favorable treatment of R&D expenditures will lead to a growth in R&D activity;
- and that growth of R&D activity will lead to more R&D "success";
- which will lead to more innovation and commercially beneficial products, which, in turn, will lead to further productivity growth.

From one point of view these assumptions do not seem troublesome. Taken individually each of the assumptions seems plausible enough and, while the empirical evidence is mixed for some propositions, the whole set of arguments seems reasonable. However, if we consider the chain of arguments as a summing of conditional probabilities, the problem appears more nettlesome. Further, if one assumption is not valid the whole logical edifice tumbles.

It is important to dissect these three assumptions in order to clarify the problematic nature of the underlying arguments. The assumption that favorable treatment of R&D expenditure will lead to a growth in R&D activity is not patent, but it can be treated briefly since the assumption and its problems are very familiar. It is sufficient to point out that R&D investment is, in general, poorly understood because of its heterogeneous nature. Morever, given our findings in Chapter 4 concerning the many various determinants of R&D, there is reason to believe that response to R&D incentives will vary considerably according to industry, firm size, current level of R&D intensity, profitability, competitiveness of markets, strategic plan, and organization composition.

The assumption that more R&D activity will lead to more R&D successes seems reasonable, although empirical evidence is not available. The more relevant question is whether the success rate will be stable. It seems plausible that certain threshold effects may operate such that there are diminishing returns from additional R&D. Illustrating this fact, Hufbauer poses an argument for a R&D scattershot approach:

> R&D expenditure is not an end in itself. The goal is successful product and process innovation. Yet expenditure is often used as a surrogate for the target, on the argument that plain luck separates R&D failures from R&D success. In this view, more R&D "shots" will increase the probability of a "hit," but nothing much can be done to shape the underlying probability distribution. (1981, p. 146)

While this kind of thinking is rarely made explicit, scattershot assumptions seem integral to many policy proposals. To be sure, the scattershot argument may be correct or, more likely, the threshold effects may be such that the point of diminishing return is distant and its slope modest. Nevertheless, the assumption that increased R&D activity will lead to corresponding increases in R&D success is one that needs more thought and should be carefully weighed.

The assumption that innovation and new technologies lead to productivity growth is not only reasonable, it is also well documented. The question here (and an open-ended question it is) is: How much does technological advance contribute to productivity growth? No one argues against the view that economic growth is steeped in complexity or that technical change accounts for only some fraction of economic growth. We have already presented evidence that technological progress and productivity are closely associated, but estimates of the magnitude of the contribution of new R&D related technology to growth vary considerably. On the one hand, Nadiri (1980) estimates that the slowdown in the stock of R&D (through the slowdown in growth of R&D spending) may account for as much as one-third of the productivity decline in the economy since 1973. On the other hand, Denison (1979) and Griliches (1980a) aver that the slowdown in R&D is not a significant factor; at best it accounts for one-tenth of the decline in productivity.

In sum, the role of R&D spending in productivity growth is extremely complex. The relation of capital formation in general to R&D capital formation in particular complicates matters. Furthermore, the effects of R&D spending according to sector are difficult to assess. The growth rate for R&D spending in government declined for several years. But industrial R&D has steadily increased. Also of some significance is the fact that the United States already has a large trade surplus in R&D intensive goods (Carlson 1981). The case for policies to stimulate additional R&D spending is hardly cut and dried.

One argument often advanced for increased support of industrial R&D, and especially tax incentives for R&D expenditures, is the claim that many of the nations that compete with the United States in international markets offer more generous tax incentives for R&D (Collins, Hammer and Randolph 1981). The CED (1980) report makes much of these differences and even cites Kaplan's (1975) study in highlighting the more favorable tax incentives of Japan, West Germany, Canada, and France. Understandably, the CED study does not quote Kaplan's conclusions that "most countries are moving away from general tax incentives for R&D" (1975, p. 51), or that "it is difficult to find disinterested observers who favor expanding the role of tax incentives in stimulating innovative activities" (p. 52). Regardless of one's position on the need for additional R&D

expenditures or further tax incentives for R&D, international comparisons offer little justification. Without information on the actual effects of the tax incentives provided by other countries, presenting a list of other countries' tax incentives as an argument for further tax incentives here is, according to Collins (1980), camouflaging the gaps in knowledge with pieces of gratuitous information.

Despite our cautious tone, we do not intend to imply that there is not a need to encourage, or even directly assist, further R&D spending. Our purpose is to point out that the role of R&D in productivity growth is complex and that one should avoid jumping quickly to conclusions. Many studies have, indeed, found a strong correlation between R&D activity and growth, but there is no easy answer to the question: How much R&D is enough?

Questions of Policy Tool Selection

Given the assumption that there is a need to stimulate R&D spending in the industrial sector of the U.S. economy, one must ask if additional tax incentives are preferable to direct assistance, subsidies, and other policy tools. There are compelling arguments for and against the use of tax incentives to achieve policy goals; most of these arguments are general but many seem appropriate to our particular interest in stimulating R&D. Before reviewing some of these general arguments, it is important to note that economists have not found it easy to estimate the impacts or costs of tax incentive programs.

One of the most important points to bear in mind about tax incentives is that they are not costless. As Cole (1971) and others have noted, there appear to be different psychological reactions to grants assistance and tax incentives. Tax incentives draw support from the rich taproot of the free enterprise symbols, but the psychological and symbolic characteristics of tax incentives must be separated from any economic analysis of impacts and costs. Some argue that tax incentives are nothing more than tax *expenditures*. According to Surrey, for example, the tax expenditure concept considers special tax incentive provisions

> as composed of two elements: the imputed tax payment that would have been made in the absence of the special provision (all else remaining the same) and the simultaneous expenditure of that payment as a direct grant to the person benefited by the special provision. (1973, p. 6)

The difficulties in gauging the impacts of tax incentives are underscored by disparities in studies of the effects of investment tax credits and accelerated depreciation (Hall and Jorgenson 1967, 1969; Bischoff 1971; Coen 1971; Klein and Taubman 1971). As Fisher (1971) points out, most of the difficulties in assessment arise from disagreements about the determinants of investment. The impact of tax incentives on R&D expenditures has been studied using models analogous to those employed in the study of the effects of tax incentives on physical investment expenditures. The problem with this approach is that R&D expenditures, unlike physical investment expenditures, are characterized by a high degree of uncertainty, significant indivisibilities, and only partial appropriability of benefits (Nadiri 1981). Nevertheless, it is likely that R&D investment is influenced by many of the same macroeconomic variables as physical investments.

Perhaps the only study designed specifically to determine the impact of various levels of a tax credit at different time horizons is by Brimmer and Company (1979). They developed an econometric model to estimate the impact of two alternative R&D tax credits, one of 10 percent and one of 27.5 percent.[7] One of the advantages of their approach is that it contains the possibilities of certain threshold effects. The study concludes that a 10 percent tax credit effective during the period 1966-76 would have had a tax cost of about $1.1 billion per year (in constant 1972 dollars) but would have increased GNP by an average of $1.92 billion per year. The larger 27.5 percent tax credit provides a greater stimulus to R&D expenditures and productivity during the same base period, but of course there is a larger tax cost. Interestingly, the R&D spending variable is less elastic in response to the 27.5 percent credit than in response to the 10 percent credit, and, while there are substantial lag effects with each credit, the lag effects are more pronounced for the larger credit.

For present purposes, the likely impact of the tax credit introduced in the Tax Act is of special interest. While estimates of the credit's impact vary, one established estimate is that of the Joint Economic Committee of the U.S. Congress. The Committee estimated that the tax credit would result in a loss of from $400 to $900 million in revenue to the treasury (Ruscio 1981). Assuming that the range of this estimate is accurate, it can be used to frame a single issue: What different results would be expected if the same amount of money had been appropriated for a direct assistance program for support of additional industrial R&D? Following this approach the apparent trade-offs between tax incentives as a policy tool and grants assistance can begin to be assessed.

Consider some of the arguments advanced in favor of tax incentives and, in particular, tax incentives for R&D expenditures.

The major arguments for tax incentives as a policy tool include the following:

- Tax incentives entail less interference in the marketplace than grants assistance and, in relation, allow private decision makers to retain their autonony.
- Tax incentives require less paperwork and there is no need for another layer of bureaucracy to administer grants.
- It is possible to avoid making difficult distinctions and setting nebulous and particularistic requirements for receiving assistance.
- There are psychological advantages in that industry simply reacts differently to tax incentives than to grants assistance or subsidies.
- There is a permanence and stability to tax incentives as they do not require annual budget review.

To these more commonly advanced arguments, one more can be added: Tax incentives have a high degree of political feasibility.

These arguments appear to be straightforward, but their significance, and in some cases their validity, are being challenged. Before examining the extensive criticisms of tax incentives, we first elaborate on their advantages. It is claimed in the CED report (1980, p. 35) that "tax incentives [for R&D expenditures] do not require that government officials make difficult, subjective judgments concerning the relative merits of various innovations and technologies," and that tax incentives do not create artificial markets since "firms are still free to design, price, and sell in response to real demand, rather than in response to government-created demand." It is difficult to find fault with the first point, at least from an empirical perspective. Tax incentives do seem to provide more discretion for private decision makers. However, it is not clear that there is any necessity for grants assistance programs to encumber private decision making since it is theoretically possible to devise grants assistance and subsidy programs that entail very few controls. By the same token, it is possible to formulate tax incentive programs that specify so many qualifying criteria that the effects are quite similar to a tightly-controlled grants assistance program. Nevertheless, it is true that tax incentives programs tend to rely more on the discretion of private decision makers and less on the preferences of government officials.

The red tape, paperwork, and additional accounting associated with many grants assistance programs are all too familiar. Tax incentives also require administration and paperwork but, according to tax incentive advocates, the administrative burden is not as great. Furthermore there is a government bureaucracy (the Internal

Revenue Service) well trained and experienced in auditing and tax administration (Cole 1971). Also, the fact that the "rules of the game" change less rapidly with tax incentives than with grants assistance programs is seen as an administrative advantage. One of the greatest boons to administrative practice is predictability, and policy flowing from tax incentives is more predictable and stable than policies, such as grants assistance programs, that require yearly appropriations and are sometimes subject to rapid legislative change.

It is argued that the greater longevity and stability of tax incentives has another advantage in respect to R&D expenditures. Firms may be more likely to make fundamental changes in their plans and investment strategies if they perceive that a policy has some stability. Thus, a tax credit that has a known long-term or open-ended effective period will be more likely to lead to adjustments in behavior than a grants assistance program or subsidy that is here today but perhaps gone with tomorrow's round of budget cutting. This seems to us one of the most powerful arguments for permanent tax incentives as a policy tool for shaping R&D activity.

The supposed psychological advantage of tax incentives vis-à-vis assistance programs is difficult to document. Cole (1971) argues that, however irrational, people do react differently to tax incentives than to direct subsidies. As mentioned above, the free enterprise ethic has considerable symbolic capacity and it would be somewhat surprising to find that there was no difference in businessmen's psychological affect for tax incentives and direct assistance, respectively. The real question is the behavioral significance of these (possibly) different reactions. If businessmen were more likely to eschew, for psychological reasons, money provided via assistance programs, this difference would be important.

Finally, political feasibility must be considered. It is often the case that tax incentives face less political opposition than direct assistance policies. What better evidence of this fact than the rapid growth of the tax expenditures budget even as officials scour budgetary lines for possible spending reductions. Also, there is a different constituency for tax incentives than for assistance programs. Tax incentives are typically the "favored child" of conservative politicians and their business constituencies not only in the United States but in other industrial nations. One of the first steps that newly elected conservative administrations take is to expand tax incentives; one of the first steps that liberal administrations take (or at least give lip service) is to close up the "loopholes." One very pragmatic argument for tax incentives for R&D expenditures is that we should not look a gift horse in the mouth. If conservative administrations choose to use this means of encouraging R&D spending, why resist?

Opponents of expanded tax incentives have levied a number of charges; some of the tax incentive advocates alleged benefits are viewed by critics as decided disadvantages. One of the harshest critics is Surrey (1973) who attempts to deflate claims of less bureaucracy and paperwork and provides a different interpretation of the effects of tax incentives on decision making. Additionally, he presents arguments that tax incentives often produce undersirable inequalities, wreak havoc with attempts to coordinate and manage policy, and produce windfalls by paying taxpayers for activities that they would have undertaken with no additional incentive.

Surrey also argues that tax incentive policies are not inherently different from grants assistance policies, but they are typically less well thought out and offer a less coordinated approach to achieving the same ends. Most of the objectives of tax incentive programs could be achieved through grants assistance programs and most of the problems encountered in grants assistance programs flow less from defects in the policy tool itself than from poor planning and administration.

One of the arguments most commonly advanced in favor of tax incentives is that they require less government interference. Surrey takes issue with this claim, contending that there is little value in comparing a direct assistance policy that is tightly controlled to a tax incentive policy that has almost no controls. If controls are the cutting issue, it is possible to devise grants assistance programs that entail almost no controls, just as it is possible to devise tightly controlled tax incentive policies. The same argument can be advanced in connection with the alleged advantages of tax incentives in promoting private decision making. Moreover, the fact that existing tax incentives are indeed less structured than most direct expenditure programs is a virtue limited to the eye of the beholder. In some instances lack of structure in tax incentives represents little more than a retreat from planning and decision-making responsibility.

If we accept critics' arguments, tax incentives offer no real advantages with respect to controls, red tape, or government interference. Perhaps even more compelling, however, are the critics' arguments for the disadvantages of tax incentive approaches. The disadvantages most commonly cited include the potential for unintended windfalls; undesirable inequities; elevation of tax rates; and the undermining of budget control and public accountability.

Critics often complain that tax incentives reward people for doing things they would have done without the tax incentive. This proposition requires, for its strict verification, knowledge that is in principle unobtainable. Once a tax incentive is actually introduced we can never know with certainty how people would have behaved had the tax incentive not been introduced. But this logical dilemma has not curtailed speculation or subsequent analysis.

In most instances windfalls are inefficiencies rather than intended consequences of policies. Windfalls can be avoided simply by introducing adequate controls and qualifications but this introduces complexities and, moreover, resurrects the very problem that tax incentives are supposed to treat – government intervention. In regard to the incentives introduced under the Tax Act, it is not yet clear that windfalls are resulting or that firms are being rewarded for actions they would have taken in the absence of an additional tax incentive. But the fact that industrial R&D has steadily increased during recent years and that R&D managers forecast more substantial growth rates in industrial R&D for both the short and intermediate term (see Chapter 3) at least gives pause.

Related to the question of windfalls is the notion that tax incentives often result in undesirable inequities. In an especially stinging indictment, Surrey contends that tax incentives are likely to introduce inequities and often mask those inequities:

> It is clear that most tax incentives have decidedly adverse effects on horizontal equity and, as respects the individual income tax, for vertical equity as well. As a consequence of these inequitable effects, many tax incentives look, and are, highly irrational if phrased as direct expenditure programs structured the same way. Indeed, it is doubtful that most of our existing tax incentives would have ever been introduced – let alone accepted – if so structured, and many would have been laughed out of Congress. (1969, p. 139)

The problem, as Paquet (1971) points out, is that tax incentives are a blunt instrument and that highly specific provisions would have to be written into the tax code to deal effectively with inequities. In many cases such highly specific rules would result in prodigious administrative problems as well as much litigation.

It is not simply that differential rewards are provided under a tax incentive – indeed differential rewards are often desirable – but there is a tendency for "upside-down benefits" (Surrey 1973). Typically, higher income taxpayers benefit more from tax incentives. Of course those at the lower end of the scale – taxpayers with no taxable income – usually do not benefit at all from tax incentives. This is especially problematic in the case of tax incentives for R&D since many new firms have no tax liability and are not profitable during their early years as they develop products and invest in R&D

assets. But in many cases those are the very firms that most often warrant assistance since they are often the most innovative and risk bearing. Carry forward provisions can in some instances begin to address this most troublesome of inequities, but even then there is potential for a "corporate regressivity" in tax incentives. After surveying the literature on the effects of tax policy on R&D, Ruscio (1981 p. 8) concludes that "the implication is that tax incentives are most effective when they are least necessary; tax incentives may influence those firms who need it least."

One of the more controversial arguments against tax incentives is that they "raid" the federal treasury with the result that tax rates are higher than they would otherwise be. Certainly tax expenditures represent foregone revenue and, other things being equal, the tax rate would have to be elevated just to maintain revenues. But of course we cannot conclude that all else is equal since revenue is not simply a function of the tax rate. Supply-side arguments such as those presented by Brimmer and Company (1979) tell us that tax expenditures can ultimately lead to an increase in revenues provided the incentives lead ultimately to productivity growth. Nevertheless, it is important to note that tax incentives are usually open-ended with no limit on the amount, only the percentage, a taxpayer can earn. There is an unpredictable element to all tax incentives since their effects are determined by a great many economic and political variables, especially general business conditions and market opportunity (Visscher 1976).

It seems clear that tax incentives for R&D provide a stabilizing influence that grants assistance programs would be hard pressed to match. Firms can, for example, have some confidence that the existing 25 percent tax credit for R&D expenditures introduced recently will be in place at least until its legislated "sunset" period. But the benefits of the stability of tax incentives are not without costs. Congress has essentially forfeited its oversight function in providing assistance through the tax credit policy tool. This has a number of ramifications. In the first place, tax incentives are awarded a status not enjoyed by other government initiatives since they in essence represent an "off budget" and uncontrollable expenditure. This not only has the effect of reducing control and coordination of policy but also shifts the locus of control from the substantive authorizing committees and the budget subcommittees – the bodies that should have the most expertise and responsibility for R&D policy – to the tax committees (the Senate Finance Committee and the House Ways and Means Committee). Congress passed the Congressional Budget and Impoundment Control Act of 1974 in part as a means to exert managerial control of the budget and eliminate "backdoor spending." Tax expenditures are an especially pernicious form of backdoor spending since the revenue committees themselves (rather than the authorizing committees) are opening the backdoor. Unless there is a clear conviction that policies implemented via tax expenditures merit an immunity not granted to other R&D assistance programs, the result is an unnecessary abrogation of policy leadership. In an era of budget cutting any step that limits the ability of Congress to manage its policy priorities should be given close scrutiny.

Questions of Policy Design

Previously, after having expressed several notes of caution, we posited the assumption that measures to increase R&D expenditures in industry are needed and we considered the case for tax incentives as the policy tool for realizing that goal. Here we assume that tax incentives for stimulating R&D are needed, and we discuss criteria for assessing tax incentives – especially tax credits for R&D expenditures. We should note that we are much impressed with some of the arguments against tax incentives and that many of our criteria are aimed at ameliorating some of the less desirable characteristics of tax incentives.

Design characteristics that seem to us, and in some instances to others (Wolfman 1965; Slitor 1966; Collins 1980; Eisner 1981; Ruscio 1981), of especial importance are: (1) effects on basic research; (2) differential treatment according to firm innovativeness and R&D success; (3) administrative ease; (4) "capacity building"; and (5) political feasibility.

Basic Research. There are theoretical reasons to believe (and to argue) that government policies should be aimed at enhancing basic research since research nearer the basic end of the spectrum is likely to provide significant social benefit. There is also some empirical evidence (presented in Chapter 5) that expenditures on basic research are highly correlated with measured productivity growth. There are fewer disincentives for industry to invest in applied research and development since the resulting benefits are more appropriable, the levels of uncertainty (as opposed to risk) are less, and there are fewer externalities. In our view, basic research should receive highest priority in any government attempt to stimulate additional R&D spending.

The Tax Act credit for incremental R&D does not fare well by this standard. While there are prohibitions against market research, demonstration projects, and quality control research, applied research can qualify as easily as basic research for the tax credit. This fact does not imply that an increase in applied research is undesirable, rather at the margin it is a less desirable target for government incentives. Of course there is a sound argument for the structuring of the R&D tax credit: it does not require difficult and perhaps unrealistic definitional distinctions between applied and basic research. Parenthetically, one of the apparent advantages of our joint venture tax credit proposal (outlined in Chapter 7) is that it does not require such explicit demarcation, but instead provides a theoretical rationale for firms to move in the direction of more basic research as a means of serving their own self-interest.

Differential Treatment. The arguments are impressive that tax incentives are often a blunt instrument and that they reward the wrong taxpayers while doing little for those that could most use the help to underwrite more innovative R&D. A particularly important consideration is that newly established firms are less likely to benefit from a tax credit of any kind. Even if there are generous carry forward provisions, the risk bearing, R&D intensive new firm may not realize any future tax benefits. This is a problem that is widely recognized and legislation is pending that would give more favorable treatment to new firms.

A major problem with tax incentives for R&D expenditures is that they fail to deal with the crucial issue of inter-industry and inter-firm differences in impact. One of the advantages of grants assistance programs is that such clear-cut distinctions can be made with little difficulty and differential rewards and incentives can be structured according to the apparent social benefits derived from assistance. It is possible to structure tax incentives so that they provide distinctions both between industries and according to firm characteristics. However, it is not a simple matter to write tax provisions that are sufficiently precise and not hopelessly confusing. Nevertheless, one important consideration for the effectiveness of tax incentives is that they have some power to benefit those institutions likely to generate the higher social returns. Likewise, benefits (especially windfalls) to unintended parties should be avoided. Otherwise tax incentives are likely to prove highly inefficient and to foster a potential for resource misallocations.

A related issue is the need to reward R&D success rather than R&D activity. Again, the bluntness of most tax incentives means that they hold little promise for distinguishing *total* R&D activity from *successful* R&D activity. Tax incentives that are sensitive to this distinction are preferable. The exception, of course, is that it is often desirable to reward new firms for R&D activity even if the R&D does not lead immediately to commercialization of technology. As Link (1980) has shown, firm size is not a prerequisite for successful R&D activities beyond some modest threshold level. Since larger firms will benefit relatively more from a tax policy, the per se thrust of a tax policy approach is less than optimal.

Administrative Ease. One of the supposed advantages of tax incentives is that they avoid the bureaucratic entanglements and red tape of many grants assistance programs. In most cases there is little art to designing a tax incentive program that is simple to administer; the problem is that administrative schemes are sometimes simple only because the policy is overly broad and insensitive to important and policy relevant differences in taxpayers. The key is to formulate a balance, and if this is not possible some other policy tool should be considered.

It is important to remember that tax incentives are not a "nonbureaucratic" approach to public policy. Often less bureaucracy is initially entailed, but sometimes the bureaucratic morass created by tax legislation is just as troublesome as even those grants assistance programs that have become legendary for their Byzantine administrative structures. As Surrey (1969, p. 131) notes, many tax policies "avoid a bureaucracy of federal experts to approve and supervise expenditures, [but] substitute tax administrators, tax planners, and a tradition of protracted administrative controversy and litigation." In sum, there is some great appeal to eliminating the need for bureaucratic rule makers, but if tax incentives simply replace one set of bureaucrats with another and one set of complex rules with another, there is little gain.

Capacity Building. Both tax incentives and grants assistance programs generate significant externalities. However, one of the traditional advantages of grants assistance programs is that those aimed at university-performed R&D have the effect of enhancing graduate education both by providing funds directly to students and supporting projects from which the students learn their trade. With the advent of indirect cost payments, grants assistance programs began to support university infrastructures and other educational and R&D assets. While there is no unanimous agreement as to the benefits of federal research assistance to the universities (Pake 1967; Pettit 1969), most observers agree that federal support of university-based research has helped build capacity. Although capacity building has sometimes been viewed as ancillary to civilian R&D programs, defense programs have been more directly concerned with capacity building.

Tax incentives can also be used for capacity building and one standard for evaluating particular proposals is whether they sow as well as they reap. There are already several provisions in the tax code that may be viewed as incentives for the private support of education and research, and the Tax Act has extended the benefits for allowable deductions for contributions of newly manufactured research equipment to universities. As universities and industries look to form closer ties, capacity building criteria should prove even more important.

Political Feasibility. The significance of political feasibility as a criterion for evaluating tax policies for R&D should be of secondary importance. It is inefficient to advance a policy on the basis of feasibility if that policy is, on balance, not desirable. However, the economic efficiency of a proposal should not be the only consideration. There is a tendency to evaluate proposals exclusively on the basis of positive economic efficiency (and even then relying on assumptions of perfect rationality that have little correspondence with either observed patterns of business decision making or citizens' preferences). As an example, though not an extreme one, consider Boskin's (1981) recommendations for changing tax policy to enhance R&D and industrial productivity. His recommendations include an integration of the corporate and personal income tax, a gradual reduction and elimination of the capital gains tax, and a gradual increase in the number of vehicles and amounts available for tax-free saving and investment. According to his plan, households would be taxed on their income minus their savings (with variations allowing for the relative wealth of households). Even if we assume that each of these proposals makes perfect economic sense, it seems clear that even in a political climate favorable to business there is virtually no chance that many of these provisions would be written into the tax code. We are not arguing that Boskin, or others, should refrain from analyzing proposals on the basis of their apparent economic rationality, only that persons considering proposals for possible mid-term political action would do well to give careful consideration to political feasibility.

CONCLUSIONS

In this chapter we have attempted to review arguments for and against stimulating R&D spending and to scrutinize the use of tax incentives as the stimulus for such spending. The overall tone was intended to provide notes of caution. The policy system for supporting R&D in the United States, in general, and in industry in particular, is inordinately complex. Nevertheless, it is interconnected. Many aspects of R&D policy are in dynamic tension, others appear to be at cross purposes, and still others appear complementary. Any attempt to provide an evaluative overview is less than complete if it focuses on only a part of the policy system, such as support for industrial R&D, and on only one policy tool, such as tax incentives. Clearly, more comprehensive treatments are needed. We would be remiss if we failed to note that piecemeal assessments, such as ours, must ultimately give way to considerations of the multiple tools and multiple purposes of R&D policy.

In Chapter 7 we argue in favor of tax incentives for joint venture R&D. Although this proposal is new, it is not without theoretical merit. We argue that this proposal overcomes many of the problems discussed in this chapter.

NOTES

1. The discussion in this section draws heavily from Slitor (1966), Kaplan (1975), Collins (1980), and Ruscio (1981).
2. An important precedent regarding the value of R&E deductions for the individual taxpayer was set in the 1974 case of *Edwin A. Snow, et. ux.*, v. *Commissioner of Internal Revenue*. By unanimous vote the Supreme Court held that a taxpayer is allowed to deduct his pro rata share of a partnership's R&D expenses under Section 174 even though the firm was not "carrying on trade or business" as required under Section 162(a) of the tax code for a trade or business expense. The Court ruled that a business is entitled to deduct R&E expenses incurred in the development of its initial product even though no product has yet been fully developed or marketed. Other codifications of Section 174 include Coors Procelain Co., 52 TC682, 1970; Charles O. Anderson (9 Cir; 2-6-79); Beaumont Co., 3 BTA822, 1970; *RevRul* 77-27, 1977; *RevRul* 77-253, 1977.
3. Since 1977, Treasury Regulation 1.861-8 required U.S. multinational firms to allocate some of their domestic R&D expenditures against income from foreign sources. The rationale is that if a firm spends money for R&D in the United States and if the products and processes are sold abroad, then a portion of these R&D costs should be allocated against foreign sales. The combined effect of the regulation and the provisions of the tax law concerning foreign income is to increase the effective tax rate on foreign income and, according to some opponents, to encourage multinational firms to move some of their R&D overseas (Ruscio 1981).
4. We demonstrated the strength of this correlation in Chapter 5. The CED report fails to point out that R&D is heterogeneous in nature and that the R&D to productivity relationship may vary depending on the character of use of R&D.

5. A more comprehensive review of the Tax Act is in Barth and Cordes (1982).
6. A further summary of these points is in Link (1981c).
7. This 27.5 percent figure is not arbitrary. According to their model, 27.5 percent is the size of the tax credit that would have been required to raise the R&D spending to sales ratio to a pre-1970 level.

7

The Economics of a Tax Credit for Joint Venture R&D

INTRODUCTION

In the previous chapter we offered several notes of caution about tax incentives for stimulating industrial R&D spending. In this chapter we propose one alternative to existing proposals. Namely, we advocate a tax credit for joint venture R&D. This proposal is chiefly motivated by the fact that current tax policies make no distinction between basic research and the other forms of activities under the R&D rubric. In other words, current policies treat R&D spending as a single homogeneous activity, and implicitly assume that all categories of R&D spending have equal impact on productivity growth.

This characterization of R&D is bothersome for at least two reasons. First, it is believed that a greater social utility is derived from marginal increases in basic research than from either applied research or development expenditures since the benefits from basic are more broadly diffused. Our finding regarding the basic research to productivity growth relationship presented in Chapter 5 lends some support to this contention. Second, the existing proposals, tax credits in particular, appear to have the disadvantage of rewarding firms for doing what they are already doing. Mansfield emphasizes this point:

> [The Economic Recovery Act tax proposal's] most important disadvantages are that they would reward firms for doing R and D that they would have done anyhow,...they would be likely to encourage the same kind of R and D that is already being done (rather than the more radical and risky work where the shortfall, if it exists, is likely to be greatest). (1981b, p. A13)

An earlier version of the material in this chapter was presented at the 1983 Annual Meeting of the American Society for Public Administration.

An obvious approach to stimulating basic research would be to provide a tax credit directly aimed at it. Such a proposal, while well intended, would be inferior to existing policies because it would require a judgment as to exactly what activities comprise basic research—such a judgment is not only difficult and costly, but probably impossible to make across all firms in all industries. Our proposal for a tax credit for joint research is in part based on a public goods model (presented subsequently) of R&D spending that suggests a tax credit may be designed to encourage firms to allocate voluntarily a large percentage of their R&D to basic research.

The possibility of rewarding firms for doing what they would have done in the absence of a tax incentive is only one of several issues in the design of a tax credit. As discussed in Chapter 6, there are characteristic advantages and disadvantages to the use of tax incentives as a policy tool. Advantages including reduction of administration, increased political feasibility, increased private sector discretion, and greater predictability of policy must be balanced against loss of accountability, possible introduction of inequities and windfalls, and loss of revenue. Tax policies are a blunt instrument and their reduced levels of targeting and control are sources of both advantage and disadvantage.

We contend that a joint venture R&D tax credit, proposed below, might well have the effect of minimizing some of the disadvantages usually associated with tax incentives while permitting the realization of the traditional advantages.

A TAX CREDIT FOR JOINT VENTURE R&D

According to a recent Conference Board report (Janger 1980), about half of the U.S. industrial firms with more than $100 million in sales are involved in joint ventures. If one also includes less formal nonequity joint ventures, one finds that most of the largest firms and a significant percentage of medium and small firms are also involved in joint ventures of one kind or another. In the latter less restrictive sense of the term, a joint venture may be defined as any form of association among two or more firms that involves collaboration and pooling of resources for more than a transitory period (Friedman and Kalmanoff 1961). Some joint ventures transcend national boundaries and involve firms from two or more countries, others do not.

Firms enter into joint ventures for any of a number of reasons including the desire to attract new markets, to procure raw materials or an export base, or to take advantage of attractive government policies in other states or nations. Many joint ventures are vehicles for the sale or licensing of technology. However, one of the most common motivations for joint ventures is to share and sometimes to diversify economic risk. The economic risks that are of interest to us are those pertaining to R&D. We shall argue that a tax credit for joint venture R&D might prove an effective means of encouraging the distribution of risks among firms and, thereby, facilitating research nearer the basic end of the basic/applied/development continuum.

Various forms of cooperative research, including formal joint ventures, are becoming more popular in this country. Japan and Europe have, of course, taken greater strides in the direction of cooperative research. In some countries, notably Japan, the government is directly involved in the planning and financing of cooperative R&D. While interest in public policies for cooperative R&D, especially for university-industry cooperative arrangements, has risen in the United States, government support is thus far negligible. It is our contention that a tax credit for joint venture R&D might prove a significant and beneficial policy innovation.

Our proposal, which at this point is not detailed, is simply that the federal government institute a tax credit for firms participating with other firms (*not* with government, universities, or not-for-profit organizations). The qualifying R&D expenditures might well be the same as with the tax credit provided under the Economic Recovery Act of 1981 (Tax Act). Furthermore, while we suggest no particular level for the credit, we do suggest that the joint venture credit be calculated in relation to tax credits claimed under the Tax Act (such that firms already receiving a full 25 percent tax credit under the Tax Act would not receive as great a credit for joint venture R&D as a firm receiving less or nothing under the Tax Act).

RATIONALE FOR A TAX CREDIT FOR JOINT VENTURE R&D

We feel that a tax credit for joint venture R&D would permit the realization of the traditional policy tool advantages of tax incentives but would be less vulnerable to typical flaws of tax incentives than those included in the Tax Act or other familiar proposals.

Like the 25 percent credit under the Tax Act, a joint venture R&D tax credit would allow for greater private sector discretion, would entail relatively modest administration costs, and would support industrial planning efforts through the stability and predictability it would provide. However, the joint venture R&D tax credit can be expected to have greater yields for research nearer the basic end of the R&D spectrum.

As we pointed out, current policies, including the R&D credit of the Tax Act, treat R&D as a single homogeneous activity and do not assume that one category (basic/applied/development) of R&D spending has a greater impact on productivity than another.

The chief advantage of a tax credit for joint venture R&D is that there are reasons to expect that a much higher percentage of joint venture R&D will be directed to the basic end of the spectrum, even in the absence of an explicit policy nudge toward basic. As we seek to demonstrate in the formal model presented in the next section, joint venture R&D will more often be directed to basic research because it is likely that it will be in the best interests of participating firms to do so. Thus, the policy is in a sense "self-executing" because it expects nothing more than that firms will act in their perceived best interests.

With or without a tax incentive stimulus, joint venture R&D tends to be a means of sharing uncertainty and distributing risk. Indeed one of the most important explanations of the increase in cooperative and joint venture research efforts is that few firms "wish to participate on an individual basis in research that is of 'collective' interest (i.e., that does not necessarily offer a competitive advantage but can raise the general level of performance of all companies in the sector)" (Rothwell 1982, p. 6).

Since basic research has more public goods characteristics than applied research or development activities, firms have less likelihood of fully appropriating the resulting knowledge and thus, a tax incentive for R&D might best be viewed by any individual firm as a means to finance research (such as development) that would have greater appropriability and would lead more directly to commercialization. The lesser degree of appropriability and the public goods characteristics of basic research would seem to make it an attractive prospect for joint venture R&D that shares risk among firms and (with the institution of a tax credit) with the government.

Also, and relatedly, we might expect that research performed in joint venture R&D would be less applied or development because all the participating firms would have an interest in deriving benefits. To the extent that there is diversity among the firms in regard to their product, there should be less interest (expecially in joint ventures involving several firms) in research that is more applied or development oriented. Even for joint ventures among firms that are homogeneous in regard to their product, the ability of other firms to benefit (through antitrust legislation) from research that is sufficiently significant to alter competition (U.S. Department of Justice 1980) should itself encourage basic research.

Cost-sharing incentives also favor basic research in joint venture R&D. We know there are financial economies of scale associated with R&D as a total, and thus it may be safe to say that, owing to the greater uncertainty of basic research, these financial economies are even greater in that kind (see Chapter 4). Therefore, cost sharing may be especially attractive for basic research activities, given the self-financing constraints firms face for R&D (and given the current economic climate).

There is yet another reason to believe that a tax credit for joint venture R&D would result in more basic research. There are significant legal constraints imposed on joint ventures (Ewing 1981). Antitrust cases could arise from joint ventures that result in market-demonstrating technologies or in restrictive agreements. Participating firms will surely take care to avoid these liabilities. One way to indemnify against the threat of antitrust action is for the firm to tilt the joint venture activity toward basic. The U.S. Department of Justice (1980, p. 3) recently took the position that "the closer [any] activity is to the basic end of the research spectrum...the more likely it is to be acceptable under the antitrust law." Their justification for this position is that the results from basic research are likely to increase the competitive environment of the industry of the joint research, which in turn will promote economic efficiency.

Thus, there are clear-cut practical arguments for a joint venture R&D tax credit. As mentioned, one of the practical arguments is a greater return to basic research. In the following section we present a graphical model that supports our contention that firms engaged in joint venture R&D find it to be in their self-interest to conduct research nearer the basic end of the R&D spectrum.[1]

A PUBLIC GOODS ANALYSIS OF R&D SPENDING

The maintained assumption of this model is that R&D is an investment in the production of technical knowledge, and that technical knowledge has the characteristic of a public good. This assumption is motivated by the belief that technical knowledge is not fully appropriated by the firm conducting the R&D, and hence will, over time, diffuse throughout the industry and the economy. Technical knowledge can either diffuse as information per se, or as information embodied in newer vintages of capital.

One model of a firm's investment in a public good is illustrated in Figure 7.1. The vertical axis measures the public good—technical knowledge. Because this knowledge is assumed to result from investments in R&D, the vertical axis can also measure the firm's resource flow or R&D expenditures. The horizontal axis measures the firm's flow of resources toward some private numeraire good.

Figure 7.1
Graphical Illustration of a Public Goods Model of R&D Spending

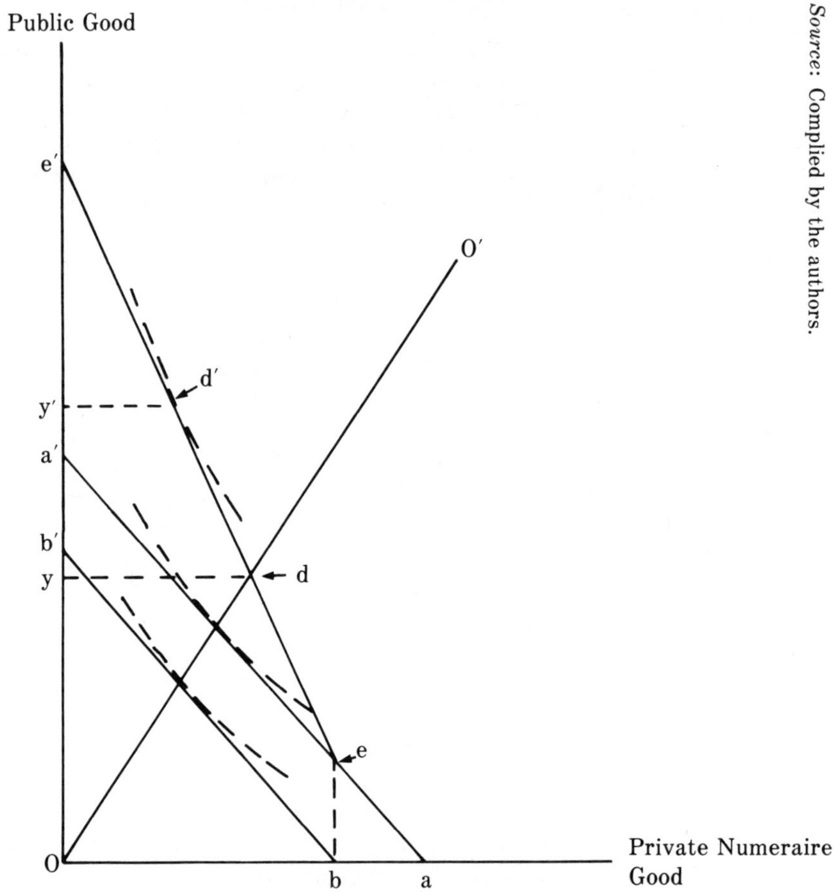

Source: Complied by the authors.

Two firms are shown in Figure 7.1. Firm A has an initial endowment of the numeraire good denoted by *Oa* and firm B has an initial endowment denoted by *Ob*. The numeraire good is transformed into the public good along transformation curves *aa′* and *bb′*. For simplicity, these transformation curves (budget lines) are assumed to have identical slopes. The ray of *OO′*s represents the locus of optimal investment choices for the two firms, acting independently.

Assume that firms A and B (within the same industry) act independently in their allocation of resources between the public and private good. By definition of public good, firms A and B receive the benefits from each other's investments (discounting any lags in the diffusion of knowledge). Since they receive equal amounts of the public good and since they allocate their resources independently of each other, each must be on ray *OO′*. Since each firm must consume equal amounts in the public good; it follows that both must reach an identical and unique allocation on *OO′*.

This solution point can be illustrated in Figure 7.1 by locating the intersection of *OO′* with the bundle of goods that firms A and B can simultaneously obtain. Clearly, firm B cannot be to the right of the point *b* on the horizontal axis. If both firms invest *Ob* in the private good, then firm A could invest the amount *be* in the public good. Therefore, point *e* represents one bundle of goods that both firms could reach simultaneously. Alternatively, if firms A and B invest *Oa′* and *Ob′* in the public good, then each could move to point *e′* ($Oe' = Oa' + Ob'$) owing to the nonexcludability of the public good. The locus of all bundles that firms A and B could reach simultaneously is *ee′*. This locus intersects *OO′* at point *d*, the independent adjustment equilibrium. The amount of resources invested by both firms at that point is *Oy*.

Point *d*, of course, is not economically efficient since both firms can be moving to point *d′*. This point can be reached only through cooperation, that is, through horizontal joint research ventures; otherwise, the free rider problem would inhibit such a move. Cooperation will increase total technical knowledge (R&D spending) between firms from *Oy* to *Oy′*.

CONCLUSIONS

We conclude that a tax credit for joint venture industrial R&D will not only increase the level of R&D but also yield social benefits that are not as easily obtained from a direct tax credit to individual firms for additional R&D spending—an increased proportion of R&D allocated to basic research.

Were such a measure introduced it might prove one of those happy instances where policy intent and individual firms' self-interest closely coincide. There are several reasons to believe that a joint venture R&D tax credit would—unlike current tax policies—stimulate more basic than applied research. While our proposed tax credit is, like other tax incentives, a blunt instrument, it is more likely than other blunt instruments to realize the specific goal of increasing basic research. But, unlike a grant assistance or more targeted tax incentive approach, it is likely to achieve this goal without requiring difficult specifications or qualifying criteria for basic research. Likewise, the administration burden produced by the proposal should be much less than would accompany a similar grant assistance program. The benefits of the tax incentive policy tool are reaped, but many of the traditional disadvantages are ameliorated or altogether avoided.

NOTES

1. This graphical analysis is based on a model of public goods first developed by Buchanan (1968) and later extended by Jeremias and Zardkoohi (1976). A more complete version of the graphical exposition is in Bozeman, Link, and Zardkoohi (1983).

8

Industrial R&D and Public Policy: Summary and Conclusions

INTRODUCTION

Thus far we have warned that our survey-based findings must be treated with caution. Our data is limited and more empirical evidence clearly is needed. However, in this concluding chapter we are somewhat less circumspect. Although this chapter provides a summary of our findings, it also presents some preliminary thoughts about possible implications for industrial R&D policy making.

The incomplete findings presented in the earlier chapters do not permit us to recommend confidently that one industry-focused public policy course is superior to another. Moreover, the state of the art in the economics of R&D is such that it alone does not serve well as a policy-making guidepost. However, despite formidable data and modeling limitations, we would like to point out some implications that are not inconsistent with our research results.

After suggesting possible implications for policy, we return in the concluding section to the limitations of R&D economics as a guide to policy making in general. We seek there to identify gaps in our knowledge of industrial R&D, focusing on the research needed to inform public policy.

DISAGGREGATED R&D: IMPLICATIONS FOR PUBLIC POLICIES

Aggregation and Broadly Bundled Policies

It is perhaps not an overstatement to say that R&D policy making is characteristically "broadly bundled." Policies for industrial R&D are encompassing and typically make little or no provision for potentially relevant differences (in behavior and, thus, policy effects) according to industry, source of technology, product type, capital intensity, and so forth. Our thesis is that policy making for industrial R&D is often broad and roughly hewn where it should be targeted and clearly specified. This observation coincides with the major thrust of our empirical findings that R&D is not a monolith and is better understood when disaggregated.

We feel that one of the keys to more effective policy making for industrial R&D lies in the ability to specify and target policies. Insensitive policies are often inefficient. Policies that are not properly specified and targeted have a greater likelihood of rewarding behaviors that need no inducement or, worse, rewarding the wrong (from a public interest standpoint) behaviors.

There are several reasons why R&D policy has been, and still is, so poorly specified. Often, there are political advantages to broad and poorly focused policies. Distributive policies are often more feasible and less painful than policies that provide more clearly for "losers" as well as "winners." For example, a tax incentive scheme that appears inefficient in that it provides resources to some with no expectation of returns in enhanced productivity or innovations may still be *politically* efficient. However, if we dismiss political and social explanations of the inefficiency and insensitivity of policy toward industrial R&D, another crucial determinant remains—ignorance. Often, R&D policy is inefficient and insensitive simply because policy makers do not have enough information about such factors as the composition of R&D, the effects of federal funding on private R&D spending, or general differences across industries to devise better specified and more efficient policies.

Unfortunately, R&D economics in its current elliptic state can do little to enlighten policy makers. Many of the issues of concern to policy makers have been little researched and many of the issues that have received attention remain unresolved. Thus, the broadly bundled nature of policy making is in large part due to the absence of sufficient policy relevant findings in R&D economics and related fields.

We do not pretend that the results reported in this slim volume are sufficiently valid or generalizable that they should guide policy. However, our findings perhaps can help to structure policy alternatives. For example, they may suggest some dimensions for the specification of R&D policies, though they give little support for any particular specification. We take our findings as a point of departure for suggesting dimensions along which R&D policy might be specified.

Policy Specification and the Heterogeneity of R&D

There is a widespread recognition that public policy for R&D might have different effects on different industries. Likewise, research suggests that factors such as firm size, concentration, and diversification are potentially significant determinants of firms' R&D activity. However, several factors that might mediate the effects of public policies for industrial R&D remain poorly understood. One such factor, about which little is known, is source of technology.

In Chapter 2, we examined empirically differences in firms' sources of technology and sought to understand circumstances that lead firms to induce technology as opposed to procuring technology from external sources (such as from other firms or the federal government) through mergers, licensing, or capital purchases. The findings we reported suggest that source of technology may be important for understanding firms' behavior and, by extension, for developing industrial R&D policy.

In examining sources of technology, our focus was on physical capital, and therefore it is important to note the limitation of our results. Clearly cognitive technology transfer is often critical (Rogers 1982), but the flow of knowledge embodied in individuals is less easily measured, especially beyond the boundaries of the organization (Bozeman and McGowan 1982; Bozeman et al. 1978). The hypothesis tested was that the propensity to induce technology is a function of the stage of process development of a firm's industry. Our proxies for the stage included capital intensity of the industry (the reasoning is that industries in a fluid stage are more labor intensive while those in a specific stage are more capital intensive) and degree of product standardization. Our hypothesis that the relation of propensity to induce to stage of process development will follow an inverted U-shape was not disconfirmed: propensity to induce was positively related to high and low capital intensity. More specifically, we found that propensity to induce is negatively related to standardization and positively related to concentration, size, diversification, and previous R&D success.

In addition to these findings about the "make versus buy" decisions, we reported in Chapter 4 results pertaining to the federal government as a source of technology. Our conclusion of complementarity indicates that the federally financed R&D has salutary effects in some cases but not others: firms' spending for R&D is positively related to federally financed R&D for those firms whose research activity is closer to the development end of the spectrum.

These results regarding source of technology, preliminary as they are, point to some questions that might (should) be entertained by policy makers. In the first place, our findings concerning the propensity to induce technology suggest that in some cases government efforts to stimulate R&D efforts may best be directed toward an industry's suppliers. Firms in industries characterized by a high degree of product standardization and producing at the specific stage of process development may have little incentive to increase R&D activities directed at innovation or improvements in product quality. So long as competition is based on efficiency and quantity of output rather than on product quality or innovation, federal efforts to stimulate innovation-directed R&D are likely to have little payoff.

If a public policy strategy aims at building on existing R&D strength—that is, focusing on industries and firms that have demonstrated a commitment to generating indigenous technology—incentives might best be directed to large, diverse firms producing nonstandardized products within a concentrated industry. Of course, a quite different policy strategy might be required for the more ambitious (and perhaps unrealistic) goal of stimulating R&D in industries and firms that have not been oriented recently toward indigenous R&D.

Our analysis on the complimentarity of federally financed R&D may be perceived by policy makers as both good news and bad news. The "good news" is the apparent relationship between R&D intensity and federally financed R&D. The findings indicate that a dollar increase in federal R&D increases company-financed R&D by nearly ten cents. In light of more pessimistic conclusions (Tilton 1973; Kendrick 1978) about the effects of federally financed R&D on the firms' R&D spending, policy making may receive some encouragement from our preliminary results. The "bad news" is that federally financed R&D is correlated with firms' spending for development-related activity. This is not necessarily bad news, but should be poorly received by those hoping that federally financed R&D will lead to increased R&D activity near the basic end of the spectrum.

One must take care not to make too much of these conclusions. It is, of course, possible that federally financed R&D allows firms to shift their own R&D spending toward development, but this is not all to the bad and, of course, this is only one of several plausible explanations for our finding. For example, one must consider that federal support for R&D, especially R&D that is energy or defense related, is often allocated toward firms that are, by design, oriented toward development activities. Also, it is not possible to determine the extent to which federally financed research, whether closer to the basic or closer to the development end of the spectrum, actually displaces industrial R&D spending. In all likelihood, displacement effects are in most instances neither total nor absent, but at some intermediate value. Determination of the degree of displacement may presently be beyond the means of R&D economists and policy makers.

In Chapter 4 we also examined the composition of R&D spending and again generated findings that may have implications for public policy. The results pertaining to firm profitability are especially noteworthy, and again underscore the need for a disaggregated view of industrial R&D. Profitability is positively related to total R&D spending but is not related to R&D directed especially toward product innovation. We suggested that this result was perhaps due to the probability that process innovations have multiple uses within the firm. It is also possible that firms that are product oriented are older, more standardized and less likely to be associated with rapidly developing technologies. In any event, policies aimed at stimulating R&D might have quite different effects in product-oriented firms (and product-oriented innovation) than in process-oriented firms (and process-oriented innovation). Thus, process-product distinctions prove yet another plausible criterion for policy specification.

Some of the more provocative findings reported pertain to the use of firms' marketing divisions as a source of R&D intelligence. To summarize, we found that firms more involved with basic research rely less often on their marketing division to guide R&D. Firms that make greater use of marketing as a source of R&D intelligence are disproportionately involved in product (rather than process) development, and these firms more often pursue an offensive strategy in performing R&D for product development.

There is no narrow-gauge policy inference to draw from these findings about the role of marketing in R&D planning, but there is a broad issue that strikes at the very core assumptions of R&D policy. Some policies in support of increasing industrial R&D seem to assume implicitly that a unitary rational actor makes the firms' decisions about the amount and direction of R&D. We have already noted the tendency to structure policy alternatives as though firms are inclined to induce R&D rather than secure it from external sources. But even if we set aside the nature of the firms' R&D decisions, one must strive to approximate the true structure of decision-making processes pertaining to R&D. It is often analytically convenient to posit a unitary rational actor but this analytic convenience is invariably purchased at the expense of reality. (When researchers seek to inform policy makers, they may have a tolerance for errors in the direction of theoretical elegance.) Policies that seek to affect R&D decisions, but that are based on an inadequate simplification of the firms' decision-making processes, are unlikely to succeed (although they may have wide support).

Our analysis of the relationship between marketing and R&D decisions barely scratched the surface. In all likelihood, R&D decisions are determined by much the same mix of factors that drive other major organization decisions. From a functional standpoint, we might expect that the marketing division will often be joined by accounting, legal counsel, operations, and sales divisions, among others, in its attempts to influence decisions about the amount and direction of R&D. Likewise, tax incentives will be submitted by firms to a broad-based evaluation that includes not only the projected tax savings but also the broader implications (for sales, marketing, quality control, production, and so forth) of any contemplated redirection of R&D activity. Policies for industrial R&D often seem to assume that R&D decision making is a single objective maximization problem. In most cases it is probably a multiple and conflicting objective "satisficing" problem.

Means and Ends in R&D Policy

We have noted that policy for industrial R&D can often be characterized as broadly bundled and untargeted. There is little consideration, at least as reflected in extant policies, of the source of technology, the composition of R&D, or the many elements that comprise firms' decision-making process for R&D.

There is another respect in which R&D policy is broadly bundled. As we noted in Chapter 6, the assumptions underpinning support of R&D are themselves rarely disaggregated. There is a chain of reasoning linking federal support of R&D to growth of R&D activity, to an increase in R&D "success," to innovation, to commercially viable products, to productivity growth. Not only is the entire set of assumptions rarely put to a comprehensive test, but worse, there is a widespread tendency (perhaps understandable in light of measurement difficulties) to mix input with output and to mix means with ends. In Chapter 5 we tried to clearly separate means and ends. While there is no comprehensive test of the causal linkage posited by policy makers and R&D economists, there is an effort to link beginning premises (such as investment in technology) to conclusions (such as productivity growth). From one standpoint, the research reported in Chapter 5 is the most relevant for public policy because it focuses not only on input and means but also on output and ends.

Our objective in Chapter 5 was to determine the apparent contribution of a number of disaggregated technology variables to productivity growth (the "ends" of R&D policy). The determinants of productivity are an especially compelling issue in light of recent declines in productivity growth and widespread disagreement about the causes of this decline. It is also an issue that has considerable importance for policy making. The curtailment of R&D spending accompanying the decline of productivity growth is often cited as one possible causal factor. Accordingly, many policy proposals are aimed at stimulating R&D spending; that is, pursuing the chain of assumptions outlined in Chapter 6.

We found it useful to distinguish between induced and purchased technology in analyzing the productivity issue. While both kinds of technology are important determinants of productivity growth, there are some indications that the greater contributor may be induced technology from basic research activity. This seems to suggest that policies supporting research nearer the basic end of the spectrum are advisable not only according to theoretical criteria but according to empirically derived findings regarding contribution to the chief "ends" of R&D policy—enhanced productivity growth. The tentative findings concerning the lesser productivity impacts of purchased technology are also noteworthy. Purchased technology has less impact on the productivity of individual firms than basic, but the social consequences of purchasing are more stark. Even if one grants that purchased technology is modified and perhaps "reinvented" (Rice and Rogers 1980) within the firm, there would appear to be some potential danger in the shift from induced to purchased technology. Quite simply, one might conclude that purchased technology contributes less to society's "technology reservoir" even if it does enhance the capabilities of the individual firm. If this conclusion (which has intuitive appeal but cannot be directly supported by our data) is valid, there are serious policy implications. Among other possibilities, policies that lead to a greater net increase in induced basic technologies are of greater social value than are policies that reward the purchase of technology. Thus, for example, tax incentive policies might do well to differentially reward the firm's technical activity according to its intensity in induced or purchased technology. Still, a policy rewarding selective purchased technologies might be superior to one rewarding development-related indigenous R&D.

The critical policy question is whether federal support of industrial R&D actually has any effect on productivity growth. Our preliminary findings suggest that there is a positive, though small, effect and that federally sponsored research nearer the basic end of the spectrum has the greater impact on productivity growth. Moreover, we conclude that the slowdown in productivity growth is related not only to the temporary slowdown in the rate of growth of industrial R&D spending, but also to the change in the composition of R&D away from both basic and product-related R&D to more short-term and applied ends. If our data in Chapter 3 projecting R&D activity are valid, the expected trends toward more basic and more long-term industrial R&D might well be expected to have positive effects on productivity growth in the near and intermediate range future.

TOWARD MORE POLICY RELEVANT R&D ECONOMICS: SOME GAPS IN OUR KNOWLEDGE

In this concluding section we briefly discuss some perceived gaps in R&D economics research, focusing on issues that seem to be important from the standpoint of providing useful guidelines for policy making.

In our view, the state of the art in R&D economics is such that the ability to provide reliable advice for policy making is a goal that is unlikely to be attained in the near future. There are so many gaps, both in theory and empirical research, that a specific listing would itself be a formidable task. However, our purpose is not to pose a specific research agenda, only to suggest a few nonexhaustive categories that might be pursued.

Organization Decision Making and R&D Economics

Perhaps the most pressing need, not only for *policy* relevant R&D economics but for *theoretically* relevant R&D economics, is for research that considers variations in organizational decision making. Most theoretical formulations treat the firm as a unitary actor and, sometimes, specify overly simple rationality assumptions. This approach effectively ignores the last several decades of development in organization behavior and decision-making research. There is a rich and abundant body of work in organization theory that could be used to redirect R&D economics. This is not to say the task is simple. Analytical models in R&D economics that paid more attention to the determinants of organization decision making (and other organization processes) would require greater complexity, specification of a greater range of initial conditions, and in many instances, linkages between levels (intra-firm, inter-firm, industry) of analysis. Just as policy response functions are sometimes built into certain economic models, it may be possible to construct firm response models that detail expected reactions to economic and technical change according to the organization structure and process variables that distinguish firms. A modest beginning, but still in the right direction, would simply consider (perhaps as intervening variables) the impact of such standard organization variables as decentralization, administrative intensity, bureaucratic control, and hierarchy as they pertain to firms' reactions to both economic and technical change.

R&D economists are not alone in this parochialism. Researchers concerned with R&D management and industrial engineering are no more likely to exploit developments in R&D economics than economists are likely to consider organization behavior research. "Middle level" research is virtually nonexistent.

Life Cycles

Much of R&D economics can be characterized as static. There is some recognition of product cycles (Vernon 1966) and industry cycles, but little research in R&D economics focuses on the life cycles of firms. Recent developments in organization research (Kimberly and Miles 1980) have demonstrated that organizations may be viewed as having life cycles and can be expected to behave differently at various stages in their existence. More broadly, recently developed "population ecology" theories of organization (Aldrich 1979) demonstrate the dynamic relationship between organizations' life cycle stages and responses to environmentally induced change. Research in R&D economics that considers the joint effects of a number of cyclical variables, such as product stage, firms' life cycle, industrial cycles, and, of course, macroeconomic cycles, is much more likely to provide reliable guidelines about the direct and indirect effects of R&D policies.

Incentives for R&D Activity and Other Productivity Inputs

There is a widespread recognition that direct tax incentives for R&D activity constitute only part of the tax incentives for the productivity picture. For example, in its recommendations for change in tax policy, the Committee for Economic Development (1980) cited the need for changes in the tax on capital gains, credit for investments in R&D facilities (in addition to incentives for wages and direct R&D expenditures), and changes in the depreciation rates for technology. More broadly, some observers feel that such diverse factors as barriers to the formation of venture capital, the need for retooling and plant revitalization, and changes in the composition of the labor force are more important explanations of productivity decline than are the direction, composition, and diminished rate of R&D activity.

It seems to us that what is needed is not a disembodied policy for industrial R&D, but a broader policy for industrial productivity of which R&D policy is only a part. Current practices in R&D economics are such that treatment of industrial R&D is, at the same time, too focused and too aggregated. This mirrors (without necessarily shaping) public policy making for industrial R&D, which is too narrow in its conception (isolating R&D) and to broad in its application (failing to make important distinctions concerning attributes of R&D activity).

Future efforts by R&D economists might be most helpful to policy makers if research can somehow link variables pertaining directly to R&D with variables representing other possible determinants of productivity growth. There is, of course, some current effort in this direction but such a conception of R&D economics is nascent and as yet poorly developed.

Disaggregated R&D

A basic conclusion from this study is that there seems to be valid theoretical as well as instrumental reason to move away from research treating R&D as homogeneous. However, efforts (including this one) to understand R&D by disaggregating it along various lines is preliminary and quite limited in nature. Future research should seek to develop more refined measures for disaggregating R&D and might well consider a variety of additional variables as candidates for disaggregation. Perhaps such effort would, in the long term, result in greater public policy specification and more efficient allocation of resources toward inventive activity.

BIBLIOGRAPHY

Abernathy, William J. 1976. "Production Process Structure and Technological Change." *Decision Sciences* 7: 607-19.

______. 1978. *The Productivity Dilemma.* Baltimore: Johns Hopkins University Press.

Abramowitz, Moses. 1956. "Resource and Output Trends in the U.S. Since 1970." *American Economic Review* 46: 5-23.

______. 1981. "Welfare Quandries and Productivity Concerns." *American Economic Review* 71: 1-17.

Adams, William J. 1970. "Firm Size and Research Activity: France and the United States." *Quarterly Journal of Economics* 84: 386-409.

Alchian, Armen A., and Harold Demsetz. 1972. "Production, Information Costs, and Economic Organization." *American Economic Review* 62: 277-95.

Aldrich, Howard. 1979. *Organizations and Environments.* Englewood Cliffs, N J :
Prentice-Hall.

Armour, Henry O., and David J. Teece. 1980. "Vertical Integration and Technological Innovation." *Review of Economics and Statistics* 62: 470-74.

Arrow, Kenneth J. 1962. "Economic Welfare and the Allocation of Resources for Invention." In *The Rate and Direction of Inventive Activity,* edited by Richard R. Nelson, pp. 609-25. Princeton: Princeton University Press.

Baily, Martin N. 1981a. "Productivity and the Services of Capital and Labor." *Brookings Papers on Economic Activity* 1: 1-65.

______. 1981b. "The Productivity Growth Slowdown and Capital Accumulation." *American Economic Review* 71: 326-31.

Baker, Norman R., and D. J. Sweeney. 1977. "An Assessment of Modeling Capability Related to the Process of Organized Technological Innovation within the Firm." Washington, D.C.: National Science Foundation report.

Barnett, H. G. 1953. *Innovation: The Basis of Cultural Change.* New York: W. W. Norton.

Barth, James R., and Joseph J. Cordes. 1982. "The Economic Recovery Act of 1981: Impacts on R&D, Technological Change, and Innovation." Mimeographed.

Baumol, William J. 1967. *Business Behavior, Value and Growth,* rev. ed. New York: Harcourt Brace & World.

Beard, Miriam. 1938. *A History of the Businessman.* New York: Macmillan.

Berle, Adolf A., and Gardiner C. Means. 1932. *The Modern Corporation and Private Property.* New York: Commerce Clearing House.

Berndt, Ernst R. 1980a. "Comment to Gollop and Jorgenson." In *New Developments in Productivity Measurement and Analysis,* edited by John W. Kendrick and Beatrice N. Vaccara, pp. 124-36. Chicago: University of Chicago Press.

______. 1980b. "Energy Price Increases and the Productivity Slowdown in the United States." In *The Decline in Productivity Growth,* pp. 60-89. Boston: Federal Reserve Bank of Boston.

Bischoff, Charles W. 1971. "The Effects of Alternative Lag Distributions." In *Tax Incentives and Capital Spending,* edited by Gary Formar, pp. 61-130. Washington. D.C.: Brookings Institution.

Blank, David M., and George J. Stigler. 1957. *The Demand and Supply of Scientific Personnel.* New York: National Bureau of Economic Research.

Boden, L. I.; M. B. Zimmerman; and D. Spiegelman. 1981. "The Effects of Mine Safety and Health Administration Enforcement on the Cost of Underground Coal Mining." Washington, D.C.: U.S. Department of Labor report.

Boskin, Michael J. 1978. "Taxation, Savings and the Rate of Interest." *Journal of Political Economy* 86: S3-S527.

______. 1981. "Taxation, Innovation and Economic Growth." In *Taxation, Technology and the U.S. Economy,* edited by Ralph Landau and N. Bruce Hannay, pp. 35-44. Elmsford, N.Y.: Pergamon Press.

Boudreaux, Kenneth J. 1973. "Managerialism and Risk-Return Performance." *Southern Economic Journal* 39: 366-73.

______. 1975. "The Profit Variability Effects of the Managerial Enterprise: Comment." *Economic Inquiry* 13: 124-26.

Bower, Richard S. 1974. "Managerialism and the Risk-Return Performance: Comment." *Southern Economic Journal* 40: 505-06.

Bozeman, Barry, and Albert N. Link. 1982. "Tax Incentives for R&D: Some Notes of Caution." Paper presented at the Fourth Annual Research Conference of the Association for Public Policy Analysis and Management, Minneapolis, Minn. Mimeographed.

______. 1983. "Getting Full Value from Foregone Dollars: The Case for a Joint Venture R&D Tax Credit." Paper presented at the 1983 Annual Meeting of the American Society for Public Administration, New York. Mimeographed.

______; Kenneth Roering; and E. Allen Slusher. 1978. "Social Structure and the Flow of Scientific Information in Public Agencies." *Research Policy* 7: 384-405.

______, and Robert McGowan. 1982. "Information Acquisition and Political Environments in Technology Intensive Government Agencies." In *The Transfer and Utilization of Technical Knowledge,* edited by Devendra Sahal, pp. 147-74. Lexington, Mass.: D. C. Heath.

_____ ; Albert N. Link; and Asghar Zardkoohi. 1983. "The Economics of a Tax Credit for Joint Venture R&D." Working Paper E830408, University of North Carolina at Greensboro. Mimeographed.

Branch, Ben. 1974. "Research and Development Activity and Profitability: A Distributed Lag Analysis." *Journal of Political Economy* 82: 999-1011.

Brimmer and Company, Inc. 1979. "Technology, Innovation, and Economic Growth." Mimeographed.

Brittain, Jack W., and John H. Freeman. 1980. "Organization Proliferation and Density Dependent Section." In *The Organizational Lifecycle*, edited by J. R. Kimberly and R. H. Miles, pp. 291-338. San Francisco: Jossey-Bass.

Brozen, Yale. 1965. "R & D Differences Among Industries." In *Economics of Research and Development*, edited by Richard A. Tybout, pp. 90-128. Columbus: Ohio State University Press.

Buchanan, James M. 1968. *The Demand and Supply of Public Goods.* Chicago: Rand McNally.

Buer, Terje K. 1980. "Investigation of Make or Buy as a Measure in Innovation Research: A Pilot Study." Cambridge, Mass.: Massachusetts Institute of Technology. Mimeographed.

Business Week. 1982. "Annual Scoreboard of R & D Spending." July 5, 1982, pp. 54-74.

Carlson, George N. 1981. "Tax Policy Toward Research and Development." In *Taxation, Technology and the U.S. Economy*, edited by Ralph Landau and N. Bruce Hannay, pp. 63-86. Elmsford, N.Y.: Pergamon Press.

Carmichael, Jeffrey. 1981. "The Effect of Mission-Oriented Public R&D Spending on Private Industry." *Journal of Finance* 36: 617-27.

Christainsen, Gregory B., and Robert H. Haveman. 1980. "The Determinants of the Decline in Measured Productivity Growth: An Evaluation." In *Special Study on Economic Change*, vol. 10, pp. 1-17. Washington, D.C.: Congress of the United States.

_____. 1981. "Public Regulations and the Slowdown in Productivity Growth." *American Economic Review* 71: 320-25.

Clark, John Bates. 1892. "Insurance and Business Profits." *Quarterly Journal of Economics* 7: 45-54.

Clark, Kim, and Zvi Griliches. 1981. "Productivity Growth and R and D at the Business Level: Preliminary Results from the PIMS Data Base." Paper presented at the National Bureau of Economic Research Conference on R and D, Patents and Productivity, Lenox, Mass. Mimeographed.

Clark, Peter K. 1981. "Inflation and Productivity Growth." Paper presented at the Third Annual Conference on Current Issues in Productivity, Rutgers University, N.J. . Mimeographed.

_____. 1982. "Inflation and the Productivity Decline." *American Economic Review* 72: 149-54.

Coase, Ronald H. 1937. "The Nature of the Firm." *Economica* New Series IV: 386-405.

Coen, Robert, M. 1971. "The Effect of Cash Flow on the Speed of Adjustment." In *Tax Incentives and Capital Spending*, edited by Gary Fromm, pp. 131-96. Washington, D.C.: Brookings Institution.

______, and Bert G. Hickman. 1980. "Investment and Growth in an Econometric Model of the United States." *American Economic Review* 70: 214-19.

Cole, Robert T. 1971. "Discussion." In *Tax Incentives*, edited by Stanley S. Surrey, pp. 74-76. Lexington, Mass.: D. C. Heath.

Collins, Eileen L. 1980. "Sorting Out the Economic Arguments Underlying Proposed Tax Incentives to Encourage Innovation." Paper presented at the American Economic Association Meetings, Denver, Colo. Mimeographed.

______, Richard M. Hammer, and M. Bradford Randolf. 1981. *Corporation Income Tax Treatment of Investment and Innovation Activities in Six Countries.* Washington, D.C.: National Science Foundation.

Comanor, William S. 1965. "Research and Technical Change in the Pharmaceutical Industry." *Review of Economics and Statistics* 47: 182-90.

______. 1967. "Market Structure, Product Differentiation, and Industrial Research." *Quarterly Journal of Economics* 81: 639-57.

Committee for Economic Development. 1980. *Stimulating Technological Progress.* Washington, D.C.: Committee for Economic Development.

Cordes, Joseph J. 1981. "Tax Policies for Encouraging Innovation: A Survey." In *Taxation, Technology and the U.S. Economy*, edited by Ralph Landau and N. Bruce Hannay, pp. 87-98. Elmsford, N. Y.:Pergamon Press.

Crandall, Robert W. 1981. "Pollution Controls and Productivity Growth in Basic Industries." In *Productivity Measurement in Regulated Industries*, edited by T. G. Cowing and R. E. Stevenson, pp. 347-68. New York: Academic Press.

Dasgupta, Partha, and Joseph E. Stiglitz. 1980a. "Industrial Structure and the Nature of Innovative Activity." *Economic Journal* 90: 266-93.

______. 1980b. "Uncertainty, Industrial Structure, and Speed of R&D." *The Bell Journal of Economics* 11: 1-28.

Dawkins, Richard. 1976. *The Selfish Gene.* New York: Oxford University Press.

Demsetz, Harold. 1969. "Information and Efficiency: Another Viewpoint." *Journal of Law and Economics* 1: 1-22.

Denison, Edward F. 1979. *Accounting for Slower Economic Growth: The United States in the 1970s.* Washington, D.C.: Brookings Institution.

Eisner, Robert. 1981. "Tax Policy for Investment: Implications for Innovation." Paper presented at The National Science Foundation Colloquium on Tax Policy and Investment in Innovation, Washington, D.C. Mimeographed.

Elliott, J. Walter. 1971. "Funds Flow vs. Expectational Theories of Research and Development Expenditures in the Firm." *Southern Economic Journal* 37: 409-22.

Ettlie, John E. 1979. "Evolution of the Productive Segment and Transportation Innovations." *Decision Sciences* 10: 399-411.

______, and David B. Vallenga. 1979. "The Adoption Time Period for Some Transportation Innovations." *Management Science* 25: 429-43.

Evans, Michael K. 1978. "Productivity – A Macroeconomic View." In *Special Study on Economic Change*, part 2, pp. 600-15. Washington, D.C.: Congress of the United States.

Ewing, K. P. 1981. "Joint Research, Antitrust, and Innovation." *Research Management* 15: 25-29.

Fabricant, Solomon. 1954. "Economic Progress and Economic Change." In *34th Annual Report of the National Bureau of Economic Research.* New York: National Bureau of Economic Research.

______. 1978. "Productivity Growth: Purpose, Process, Prospects, and Policy." In *Special Study on Economic Change*, part 2, pp. 501-31. Washington, D.C.: Congress of the United States.

Fisher, Franklin M. 1971. "Discussion." In *Tax Incentives and Capital Spending*, edited by Gary Fromm, pp. 243-55. Washington, D.C.: Brookings Institution.

______, and Peter Temin. 1973. "Returns to Scale in Research and Development: What Does The Schumpeterian Hypothesis Imply?" *Journal of Political Economy* 81: 56-70.

Fischer, William A., and Guy Black. 1979. "Federal Funding of Industrial R&D: Stimulus or Substitute?" *Research Management* 22: 27-30.

Fraumeni, Barbara M., and Dale W. Jorgenson. 1981. "Capital Formation and U.S. Productivity Growth, 1948-1976." In *Productivity Analysis: A Range of Perspectives*, edited by Ali Dogramaci, pp. 49-70. Boston: Martinus Nijhoff.

Freeman, Christopher. 1977. "Economics of Research and Development." In *Science, Technology and Society: A Cross-Disciplinary Perspective*, edited by Ina Spiegal-Rosing and Derek de Solla Price, pp. 223-75. Beverly Hills, Calif.: Sage Publications.

Freeman, Richard B., and James L. Medoff. 1979a. "New Estimates of Private Sector Unionism in the United States." *Industrial and Labor Relations Review* 32: 143-74.

______. 1979b. "The Two Faces of Unionism." *The Public Interest* 57: 69-93.

Friedman, W. G., and George Kalmanoff. 1961. *Joint International Business Ventures.* New York: Columbia University Press.

Gilpin, Robert. 1975. Prepared testimony. In *Technology and Economic Growth*, pp. 17-26. Washington, D.C.: Congress of the United States.

Globerman, Stephen. 1973. "Market Structure and R&D in Canadian Manufacturing Industries." *Quarterly Review of Economics and Business* 13: 59-67.

Goldberg, Lawrence. 1978. "Federal Policies Affecting Industrial Research and Development." Paper presented at the Southern Economic Association Meetings, Washington, D.C. Mimeographed.

Gollop, Frank M., and Mark J. Roberts. 1981. "Imported Intermediate Input: Its Impact on Sectoral Productivity in U.S. Manufacturing." In *Aggregate and Industry-Level Productivity Analyses*, edited by Ali Dogramaci and Nabil R. Adam, pp. 149-86. Boston: Martinus Nijhoff.

Gordon, Robert A. 1945. *Business Leadership in the Large Corporation.* Washington, D.C.: Brookings Institution.

Gordon, Robert J. 1979. "The 'End-of-Expansion' Phenomenon in Short-Run Productivity Behavior." *Brookings Papers on Economic Activity* 2: 447-60.

Grabowski, Henry G. 1968. "The Determinants of Industrial Research and Development: A Study of the Chemical, Drug, and Petroleum Industries." *Journal of Political Economy* 76: 292-306.

______, and Dennis C. Mueller. 1970. "Industrial Organization: The Role and Contribution of Econometrics." *American Economic Review* 70: 101-04.

Greer, Douglas F. 1971. "Advertising and Market Concentration." *Southern Economic Journal* 38: 19-32.

Griliches, Zvi. 1973. "Research Expenditures and Growth Accounting." In *Science and Technology in Economic Growth*, edited by B. R. Williams, pp. 59-83. New York: John Wiley & Sons.

______. 1979. "Issues in Assessing the Contribution of Research and Development to Productivity Growth." *The Bell Journal of Economics* 10: 92-116.

______. 1980a. "R and D and The Productivity Slowdown." *American Economic Review* 70: 343-48.

______. 1980b. "Returns to Research and Development Expenditures in the Private Sector." In *New Developments in Productivity Measurement and Analysis*, edited by John W. Kendrick and Beatrice N. Vaccara, pp. 419-54. Chicago: University of Chicago Press.

______, and Frank Lichtenberg. 1981. "R and D and Productivity Growth at the Industrial Level: Is There Still a Relationship?" Paper presented at the National Bureau of Economic Research Conference on R and D, Patents and Productivity, Lenox, Mass. Mimeographed.

_____, and Jacques Mairesse. 1981. "Productivity and R and D at the Firm Level." Paper presented at the NBER Conference on R and D, Patents and Productivity, Lenox, Mass. Mimeographed.

Hage, Jerald. 1980. *Theories of Organizations: Form, Process, and Transformation*. New York: John Wiley & Sons.

Hall, Robert E., and Dale W. Jorgenson. 1967. "Tax Policy and Investment Behavior." *American Economic Review* 57: 391-414.

_____, 1969. "Tax Policy and Investment Behavior: Reply and Further Results." *American Economic Review* 59: 388-401.

Hamberg, Daniel. 1964. "Size of Firm, Oligopoly, and Research." *Canadian Journal of Economics and Political Science* 30: 62-75.

_____. 1966. *R&D: Essays on the Economics of Research and Development.* New York: Random House.

Harrison, Bennett, and Barry Bluestone. 1982. *The Deindustrialization of America*. New York: Basic Books.

Hayes, R. H., and W. J. Abernathy. 1980. "Managing Our Way to Economic Decline." *Harvard Business Review* 58: 67-77.

_____, and S. G. Wheelwright. 1979. "The Dynamics of Process-Product Life Cycles." *Harvard Business Review* 57: 127-36.

Hébert, Robert F., and Albert N. Link. 1982. *The Entrepreneur: Mainstream Views and Radical Critiques*. New York: Praeger.

Higgins, Richard S., and Albert N. Link. 1981. "Federal Support of Technological Growth in Industry: Some Evidence of Crowding-Out." *IEEE Transactions on Engineering Management* EM-28: 86-88.

Hirsch, Barry, T. and Albert N. Link. Forthcoming. "Unions, Productivity and Productivity Growth." *Journal of Labor Research.*

Holl, Peter. 1975. "Effect of Control Type on the Performance of the Firm in the U.K." *Journal of Industrial Economics* 23: 257-71.

Horowitz, Ira. 1961. "Estimating Changes in the Research Budget." *The Journal of Industrial Engineering* 12: 114-18.

_____. 1962. "Firm Size and Research Activity." *Southern Economic Journal* 28: 298-301.

Hu, S.C. 1973. "On the Incentive to Invent: A Clarificatory Note." *Journal of Law and Economics* 16: 169-77.

Hudson, Edward A., and Dale W. Jorgenson. 1978a. "Energy Policy and U.S. Economic Growth." *American Economic Review* 68: 118-23.

_____, 1978b. "Energy Prices and the U.S. Economy, 1972-1976." *Natural Resources Journal* 18: 887-97.

Hufbauer, Gary C. 1981. "Applying Tax Policy to Generate and Implement Technology." In *Taxation, Technology and the U.S. Economy*, edited by Ralph Landau and N. Bruce Hannay, pp. 141-50. Elmsford, N.Y.: Pergamon Press.

Janger, A. R. 1980. *Organization of International Joint Ventures*. New York: Conference Board.

Jarrett, J. Peter, and Jack G. Selody. 1982. "The Productivity-Inflation Nexus in Canada." *Review of Economics and Statistics* 64: 361-67.

Jeremias, R., and A. Zardkoohi. 1976. "Distributional Implications of Independent Adjustments in an Economy with Public Goods." *Economic Inquiry* 14: 305-08.

Johannisson, B., and C. Lindstrom. 1971. "Firm Size and Inventive Activity." *Swedish Journal of Economics* 73: 427-42.

Kamien, Morton I., and Nancy L. Schwartz. 1970. "Market Structure, Elasticity of Demand and Incentive to Invent." *Journal of Law and Economics* 13: 241-52.

______, 1976. "On the Degree of Rivalry for Maximum Innovative Activity." *Quarterly Journal of Economics* 90: 245-60.

______, 1978. "Self-Financing of an R&D Project." *American Economic Review* 68: 252-61.

______, 1982. *Market Structure and Innovation*. Cambridge: Cambridge University Press.

Kania, J. J., and J. R. McKean. 1976. "Ownership, Control, and the Contemporary Corporation: A General Behavior Analysis." *Kyklos* 29: 272-91.

Kaplan, Richard S. 1975. *Tax Policies for R&D and Technological Innovation: Tax Policies of U.S. and Foreign Nations in Support of R&D and Innovation*. Washington, D.C.: National Technical Information Service.

Kendrick, John W. 1978. Prepared Testimony. In *Special Study on Economic Change*, part 2, pp. 622-37. Washington, D.C.: Congress of the United States.

______. 1980a. "Productivity Trends in the United States." In *Lagging Productivity Growth: Causes and Remedies*, edited by Shlomo Maital and Noah Meltz, pp. 9-30. Cambridge, Mass.: Ballinger.

______. 1980b. "Remedies for the Productivity Slowdown in the United States." In *Lagging Productivity Growth: Causes and Remedies*, edited by Shlomo Maital and Noah Meltz, pp. 247-63. Cambridge, Mass.: Ballinger.

______. 1980c. "Survey of the Factors Contributing to the Decline in U.S. Productivity Growth." In *The Decline in Productivity Growth*, pp. 1-21. Boston: Federal Reserve Bank of Boston.

______, and Elliot S. Grossman. 1980. *Productivity in the United States: Trends and Cycles*. Baltimore: Johns Hopkins University Press.

Kimberly, John R., and R. Miles. 1980. *The Organizational Life Cycle*. San Francisco: Jossey-Bass.

Klein, Burton, H. 1979. "The Slowdown in Productivity Advances: A Dynamic Explanation." In *Technological Innovation for a Dynamic Economy*, edited by Christopher T. Hill and James M. Utterback, pp. 66-117. New York: Pergamon Press.

______. 1981. "A tax Policy to Internalize Risks." In *Taxation, Technology and the U.S. Economy*, edited by Ralph Landau and N. Bruce Hannay, pp. 173-204. Elmsford, N.Y.: Pergamon Press.

Klein, Lawrence, L., and Paul Taubman. 1971. "Estimating Effects Within A Complete Econometric Model." In *Tax Incentives and Capital Spending*, edited by Gary Fromm, pp. 197-242. Washington, D.C.: Brookings Institution.

Klotz, Benjamin; Rey Madoo; and Reed Hansen. 1980. "A Study of High and Low 'Labor Productivity' Establishments in U.S. Manufacturing." In *New Developments in Productivity Measurement and Analysis*, edited by John W. Kendrick and Beatrice N. Vaccara, pp. 239-86. Chicago: University of Chicago Press.

Kopcke, Richard W. 1980. "Capital Accumulation and Potential Growth." In *The Decline in Productivity Growth*, pp. 26-53. Boston: Federal Reserve Bank of Boston.

Larner, Robert J. 1970. *Management Control and the Large Corporation.* New York: Dunellen.

Lee, L. Douglas. 1980. "Productivity, Inflation, and Economic Growth." In *Special Study on Economic Change*, vol. 10, pp. 46-57. Washington, D.C.: Congress of the United States.

Lee, Tom, and Louis L. Wilde. 1980. "Market Structure and Innovation: A Reformulation." *Quarterly Journal of Economics* 94: 429-36.

Leonard, William N. 1971. "Research and Development in Industrial Growth." *Journal of Political Economy* 79: 232-56.

Levin, Richard C., and Peter Reiss. 1982. "Tests of a Schumpeterian Model of R and D and Market Structure." Paper presented at the National Bureau of Economic Research Conference on R and D, Patents and Productivity, Lenox, Mass. Mimeographed.

Levy, David M., and Nestor E. Terleckyj. 1983. "Effects of Government R&D on Private R&D Investment and Productivity: A Macroeconomic Analysis." *The Bell Journal of Economics*, forthcoming.

Link, Albert N. 1977. "On the Efficiency of Federal R&D Spending: A Public Choice Approach." *Public Choice* 31: 129-33.

______. 1978. "Rates of Induced Technology from Investments in Research and Development." *Southern Economic Journal* 45: 370-79.

———. 1980. "Firm Size and Efficient Entrepreneurial Activity." *Journal of Political Economy* 88: 771-82.

———. 1981a. "Allocating R&D Resources: A Study of the Determinants of R&D by Character of Use." Washington, D.C.: National Science Foundation report.

———. 1981b. "Basic Research and Productivity Increase in Manufacturing: Additional Evidence." *American Economic Review* 71: 1111-12.

———. 1981c. *Research and Development Activity in U.S. Manufacturing.* New York: Praeger.

———. 1982a. "A Disaggregated Analysis of Industrial R&D: Product versus Process Innovation." In *The Transfer and Utilization of Technical Knowledge*, edited by Devendra Sahal, pp. 45-63. Lexington, Mass.: D.C. Heath.

———. 1982b. "An Analysis of the Composition of R&D Spending." *Southern Economic Journal* 49: 342-49.

———. 1982c. "Alternative Sources of Technology: An Analysis of Induced Innovations." *Managerial and Decision Economics*, forthcoming.

———. 1982d. "The Impact of Federal Research and Development Spending on Productivity." *IEEE Transactions on Engineering Management* EM-29: 166-69.

———. 1982e. "Market Structure, Innovative Activity, and Voluntary Product Standards." Washington, D.C.: U.S. National Bureau of Standards policy report.

———. 1982f. "Productivity Growth, Environmental Regulations and the Composition of R&D." *The Bell Journal of Economics* 13: 548-54.

———. 1983a. "Inter-Firm Technology Flows and Productivity Growth." *Economics Letters* 11: 179-84.

———. 1983b. "Market Structure and Voluntary Product Standards." *Applied Economics* 15: forthcoming.

———. 1983c. *Measurement and Analysis of Productivity Growth: A Synthesis of Thought.* Washington, D.C.: U. S. National Bureau of Standards Monograph.

———, and Stephen O. Morrell. 1980. "Political Competition for Government Funds: Research and Development Obligations." *Public Finance Quarterly* 8: 57-67.

———, and James E. Long. 1981. "The Simple Economics of Basic Scientific Research: A Test of Nelson's Diversification Hypothesis." *Journal of Industrial Economics* 30: 105-09.

———; Gregory Tassey; and Robert W. Zmud. 1983. "The Induce versus Purchase Decision: An Empirical Analysis of Industrial R&D." *Decision Sciences* 14: 46-61.

Loeb, Peter D., and Vincent Lin. 1977. "Research and Development in the Pharmaceutical Industry: A Specification Error Approach." *Journal of Industrial Economics* 26: 45-51.

Loury, Glenn C. 1979. "Market Structure and Innovation." *Quarterly Journal of Economics* 93: 395-410.

Loveless, Steve, and Barry Bozeman. 1983. "Innovation and the Public Manager." In *Handbook of Organization Management*, edited by William Eddy, pp. 391-413. New York: Marcel Dekker.

Lunn, John. 1982. "Research and Development and the Schumpeterian Hypothesis: Alternative Approach." *Southern Economic Journal* 49: 209-17.

Manners, George E. Jr., and Howard K. Nason. 1978. "The Decline in Industrial Research — Causes and Cures." *Research Management* 24: 8-11.

Mansfield, Edwin. 1965. "Rates of Return from Industrial Research and Development." *American Economic Review* 55: 310-22.

______. 1968. *Industrial Research and Technological Innovation.* New York: W. W. Norton.

______, et. al. 1977. *The Production and Application of New Industrial Technology.* New York: W. W. Norton.

______. 1980. "Basic Research and Productivity Increase in Manufacturing." *American Economic Review* 70: 863-73.

______. 1981a. "Composition of R and D Expenditures: Relationship to Size of Firm, Concentration, and Innovative Output." *Review of Economics and Statistics* 63: 610-14.

______. 1981b. "Tax Policy and Innovation: Provisions, Proposals, and Needed Research." Paper presented at the National Science Foundation Colloquium on Tax Policy and Investment in Innovation in Washington, D.C. Mimeographed.

______, and Lorne Switzer. 1982. "Effects of Federal Support on Company-Financed R&D: The Case of Energy." Paper presented at the Southern Economic Association Meetings, Atlanta, Ga. Mimeographed.

Marcus, A. A. 1981. "Policy Uncertainty and Technological Innovation." *Academy of Management Review* 6: 443-48.

Mark, Jerome A. 1978. "Productivity Trends and Prospects." In *Special Study on Economic Change*, part 2, pp. 480-86. Washington, D.C.: Congress of the United States.

Markham, Jesse W. 1965. "Market Structure, Business Conduct, and Innovation." *American Economic Review* 55: 323-32.

Martin, Stephen. 1979. "Advertising, Concentration, and Profitability: The Simultaneity Problem." *The Bell Journal of Economics* 10: 639-47.

McEachern, William A. 1975. *Managerial Control and Performance.* Lexington, Mass.: D. C. Heath.

______, and Anthony A. Romeo. 1978. "Stockholder Control, Uncertainty and the Allocation of Resources to Research and Development." *Journal of Industrial Economics* 26: 349-61.

McGraw-Hill. 1978. "23rd Annual McGraw-Hill Survey of Business' Plans for Research and Development Expenditures, 1978-81." Mimeographed.

———. 1981. "26th Annual McGraw-Hill Survey of Business' Plans for Research and Development Expenditures, 1981-1984." Mimeographed.

———. 1982. "27th Annual McGraw-Hill Survey of Business' Plans for Research and Development Expenditures, 1982-1983." Mimeographed.

Meyer, J. R., and E. Kuh. 1957. *The Investment Decision: An Empirical Study*. Cambridge, Mass.: Harvard University Press.

Minasian, Jora R. 1962. "The Economics of Research and Development." In *The Rate and Direction of Inventive Activity*, edited by Richard R. Nelson, pp. 93-143. Princeton: Princeton University Press.

Mohr, L. B. 1969. "Determinants of Innovation in Organizations." *American Political Science Review* 63: 111-26.

Mohr, Michael F. 1980. "The Long-Term Structure of Production, Factor Demand and Factor Productivity in U.S. Manufacturing Industries." In *New Developments in Productivity Measurement and Analysis*, edited by John W. Kendrick and Beatrice N. Vaccara, pp. 137-229. Chicago: University of Chicago Press.

Monsen, R. Joseph, Jr., and Anthony Downs. 1965. "A Theory of Large Managerial Firms." *Journal of Political Economy* 73: 221-36.

Moss, Milton. 1979. "Welfare Dimensions of Productivity Measurement." In *Measurement and Interpretation of Productivity*, pp. 276-308. Washington, D.C.: National Academy of Sciences.

Mowery, David C. 1981. "The Nature of the Firm and the Organization of Research: An Investigation of the Relationship Between Contract and In-House Research." Case Western University, Cleveland, Ohio. Mimeographed.

Mueller, Dennis C. 1967. "The Firm Decision Process: An Econometric Investigation." *Quarterly Journal of Economics* 81: 58-87.

Mueller, Willard F., and Richard T. Rogers. 1980. "The Role of Advertising in Changing Concentration of Manufacturing Industries." *Review of Economics and Statistics* 62: 89-96.

Myers, John G., and Leonard Nakamura. 1979. "Data Adequacy for Productivity Analysis: A Case Study of the Primary Paper Industry." In *Measurement and Interpretation of Productivity*, pp. 391-442. Washington, D.C.: National Academy of Sciences.

Nadiri, M. Ishaq. 1980. "Sectoral Productivity Slowdown." *American Economic Review* 70: 349-52.

———. 1981. "Tax Incentives and R&D Expenditures." Paper presented at the National Science Foundation Colloquium on Tax Policy and Investment in Innovation, Washington, D.C. Mimeographed.

Nason, Howard K.; Joseph A. Steger; and George E. Manners. 1978. *Support of Basic Research by Industry.* Washington, D.C.: National Science Foundation.

National Academy of Engineering. 1980. *The Impact of Tax and Financial Regulatory Policies on Industrial Innovation*. Washington, D.C.: National Academy of Sciences.

National Research Council. 1979. *Measurement and Interpretation of Productivity*. Washington, D.C.: National Academy of Sciences.

National Science Foundation. 1982a. *National Patterns of Science and Technology Resources, 1982*. Washington, D.C.: Government Printing Office.

———. 1982b. *Research and Development in Industry, 1980*. Washington, D.C.: Government Printing Office.

Nelson, Richard R. 1959. "The Simple Economics of Basic Scientific Research." *Journal of Political Economy* 67: 297-306.

———. 1982. "The Role of Knowledge in R&D Efficiency." *Quarterly Journal of Economics* 97: 453-70.

———, and Sidney G. Winter. 1977. "In Search of a Useful Theory of Innovation." *Research Policy* 6: 36-76.

Nordhaus, William D. 1972. "The Recent Productivity Slowdown." *Brookings Papers on Economic Activity* 3: 493-536.

———. 1980. "Policy Responses to the Productivity Slowdown." In *The Decline in Productivity Growth*, pp. 147-72. Boston: Federal Reserve Bank of Boston.

Norsworthy, John R., and Michael J. Harper. 1981. "The Role of Capital Formation in the Recent Slowdown in Productivity Growth." In *Aggregate and Industry-Level Productivity Analyses*, edited by Ali Dogramaci and Nabil R. Adam, pp. 122-48. Boston: Martinus Nijhoff.

———, and Kent Kunze. 1979. "The Slowdown in Productivity Growth: An Analysis of Some Contributing Factors." *Brookings Papers on Economic Activity* 2: 387-421.

Pake, G. 1967. "Basic Research and the Financial Crisis in the Universities." *Science* 157: 517-20.

Paquet, G. 1971. "Taxation and Science Policy." *Canadian Tax Journal* 14: 429-37.

Pavitt, Keith, and Roy Rothwell. 1976. "A Comment on 'A Dynamic Model of Process and Product Innovation'." *Omega, The International Journal of Management Science* 4: 375-77.

Perry, George L. 1971. "Labor Force Structure, Potential Output, and Productivity." *Brookings Papers on Economic Activity* 3: 533-65.

———. 1977. "Potential Output and Productivity." *Brookings Papers on Economic Activity* 1: 11-47.

Pettit, L. K. 1969. "Congress, Confusion and Indirect Costs." *Science* 158: 1301-05.

Phillips, Almarin. 1966. "Patents, Potential Competition, and Technical Progress." *American Economic Review* 56: 301-10.

Phlips, Louis. 1971. "Research." In *Effects of Industrial Concentration: A Cross Section Analysis for the Common Market*, pp. 119-42. Amsterdam: North-Holland.

Press, F. 1982. "Rethinking Science Policy." *Science* 218: 28-31.

Priest, W. C. 1981. "Methodology for a Microeconomic Case Study of Productivity and Regulation: The Asbestos Industry." Paper presented at the Association for Public Policy Analysis Meetings, Washington, D.C. Mimeographed.

Qualls, P. David. 1976. "Market Structure and Managerial Behavior." In *Essays on Industrial Organization in Honor of Joe S. Bain*, edited by Richard T. Masson and P. David Qualls, pp. 89-104. Cambridge, Mass.: Ballinger.

Rasche, Robert H., and John A. Tatom. 1977a. "The Effects of the New Energy Regime on Economic Capacity, Production, and Prices." *Federal Reserve Bank of St. Louis Review* 59: 2-12.

______, 1977b. "Energy Resources and Potential GNP." *Federal Reserve Bank of St. Louis Review* 59: 10-21.

Rees, Albert. 1980. "On Interpreting Productivity Change." In *Lagging Productivity Growth: Causes and Remedies*, edited by Shlomo Maital and Noah Meltz, pp. 1-6. Cambridge, Mass.: Ballinger.

Rice, R. E., and E. M. Rogers. 1980. "Reinvention in the Innovation Process." *Knowledge* 1: 499-514.

Rogers, Everett, M. 1982. "Technological Innovation in High Technology Industries." In *The Transfer and Utilization of Technical Knowledge*, edited by Devendra Sahal, pp. 105-24. Lexington, Mass.: D. C. Heath.

______, and F. Shoemaker. 1971. *Communications of Innovations: A Cross Cultural Approach*, 2nd ed. New York: Free Press of Glencoe.

Rosenberg, Joel. 1976. "Research and Market Share." *Journal of Industrial Economics* 25: 101-12.

Rothwell, Roy. 1981. "Some Indirect Impacts of Government Regulations on Industrial Innovation in the United States." *Technological Forecasting and Social Change* 19: 57-80.

______. 1982. "Government Innovation Policy." *Technological Forecasting and Social Change* 22: 3-30.

Ruscio, Kenneth. 1981. "Tax Incentives and Innovation." Mimeographed.

Sahal, Devendra. 1981. "Alternative Conceptions of Technology." *Research Policy* 10: 2-24.

Salamon, L. 1981. "Rethinking Public Management." *Public Policy* 14: 381-92.

Scherer, F. M. 1965a. "Firm Size, Market Structure, Opportunity, and the Output of Patented Innovations." *American Economic Review* 55: 1097-125.

______. 1965b. "Size of Firm, Oligopoly, and Research: A Comment." *Canadian Journal of Economics* 31: 256-66.

______. 1967a. "Market Structure and the Employment of Scientists and Engineers." *American Economic Review* 57: 524-31.

______. 1967b. "Research and Development Resource Allocation Under Rivalry." *Quarterly Journal of Economics* 81: 359-94.

______. 1982. "Inter-Industry Technology Flows and Productivity Growth." *Review of Economics and Statistics* 64: 627-34.

Schmookler, Jacob. 1952. "The Changing Efficiency of the American Economy, 1869-1938." *Review of Economics and Statistics* 34: 214-31.

Schultz, Theodore, W. 1975. "The Value of the Ability to Deal with Disequilibria." *Journal of Economic Literature* 13: 827-46.

______. 1980. "Investment in Entrepreneurial Ability." *Scandanavian Journal of Economics* 82: 437-48.

Schumpeter, Joseph A. 1947. *Capitalism, Socialism and Democracy*, 2nd ed. New York: Harper & Bros.

Scott, John T. 1978. "Nonprice Competition in Banking Markets." *Southern Economic Journal* 44: 594-605.

______. 1981. "Firm versus Industry Variability in R&D Intensity." Paper presented at the National Bureau of Economic Research Conference on R and D, Patents and Productivity, Lenox, Mass. Mimeographed.

Shrieves, Ronald E. 1978. "Market Structure and Innovation: A New Perspective." *Journal of Industrial Economics* 26: 329-46.

Siegel, Robin. 1979. "Why Has Productivity Slowed Down?" *Data Resources U.S. Review*: 59-65.

Slitor, Richard. 1966. "The Tax Treatment of Research and Innovative Investment." *American Economic Review* 56: 217-31.

Smyth, D. J.; J. M. Samuels; and J. Tzoannos. 1972. "Patents, Profitability, Liquidity and Firm Size." *Applied Economics* 4: 77-86.

Soete, L. L. C. 1979. "Firm Size and Inventive Activity." *European Economic Review* 12: 319-40.

Solow, Robert M. 1957. "Technical Change and the Aggregate Production Function." *Review of Economics and Statistics* 39: 312-20.

Stano, Miron. 1976. "Monopoly Power, Ownership Control, and Corporate Performance." *The Bell Journal of Economics* 7: 672-79.

Stein, Herbert. 1981. "Economic Problems of The Industrial Democracies." *The AEI Economist*: 8-12.

Stokes, H. Kemble, Jr. 1981. "An Examination of The Productivity Decline in The Construction Industry." *Review of Economics and Statistics* 63: 495-502.

Strickland, Allyn D., and Leonard W. Weiss. 1976. "Advertising, Concentration, and Price-Cost Margins." *Journal of Political Economy* 84: 1109-21.

Surrey, S. 1969. "Tax Incentives: Conceptual Criteria for Indentification and Comparison with Direct Government Expenditures." *Proceedings of the Tax Institute of America*: 20-21.

______. 1973. *Pathways to Tax Reform: The Concept of Tax Expenditures.* Cambridge, Mass.: Harvard University Press.

Teece, David J. 1980. "Economies of Scope and the Scope of the Enterprise." *Journal of Economic Behavior and Organization* 1: 223-47.

______. 1981. "Market in Microcosm: Some Efficiency Properties of Vertical Integration." Stanford University, Calif. Mimeographed.

Terleckyj, Nestor E. 1960. "Sources of Productivity Advance: A Pilot Study of Manufacturing Industries, 1899-1953." Ph.D. dissertation, Columbia University.

______. 1974. *Effects of R and D on the Productivity Growth of Industries.* Washington, D.C.: National Planning Association.

______. 1980. "Direct and Indirect Effects of Industrial Research and Development on the Productivity Growth of Industries." In *New Developments in Productivity Measurement and Analysis*, edited by John W. Kendrick and Beatrice N. Vaccara, pp. 359-77. Chicago: University of Chicago Press.

______. 1982. "R and D and the U.S. Industrial Productivity in the 1970s." In *The Transfer and Utilization of Technical Knowledge*, edited by Devendra Sahal, pp. 63-100. Lexington, Mass.: D. C. Heath.

Thomas, Eleanor C. 1980. "Recent Research on R&D and Productivity Growth: A Changing Relationship Between Input and Impact Indicators?" Paper presented at the Third Conference on Science and on the Measurement of Technology Indicators, organization for Economic Cooperation and Development Headquarters, Paris. Mimeographed.

Thompson, James D. 1963. *Organizations in Action.* New York: McGraw-Hill.

Tilton, John E. 1973. "Research and Development in Industrial Growth: A Comment." *Journal of Political Economy* 81: 1245-48.

U.S. Department of Commerce, Bureau of the Census. 1972. *Census of Manufactures.* Washington, D.C.: Government Printing Office.

______. 1978. *Annual Survey of Manufacturers.* Washington, D.C.: Government Printing Office.

U.S. Department of Commerce, National Bureau of Standards. 1977. *Tabulation of Voluntary Standards and Certification Programs for Consumer Products.* Washington, D.C.: Government Printing Office.

U.S. Department of Justice. 1980. *Antitrust Guide Concerning Research Joint Ventures.* Washington, D.C.: Government Printing Office.

Utterback, James M. 1975. "Successful Industrial Innovations: A Multivariate Analysis." *Decision Sciences* 6: 65-77.

______, and William J. Abernathy. 1975. "A Dynamic Model of Process and Product Innovation." *Omega, The International Journal of Management Science* 3: 639-56.

Vernon, Raymond. 1966. "International Investment and International Trade in the Product Life Cycle." *Quarterly Journal of Economics* 80: 190-207.

Villard, Henry H. 1958. "Competition, Oligopoly and Research." *Journal of Political Economy* 66: 483-97.

Visscher, M. 1976. *Tax Policy for R&D and Technological Innovation The Effect of Tax Incentives on Investment Behavior.* Washington, D.C.: National Technical Information Service.

Wachter, M. L., and S. M. Wachter. 1981. *Toward a New U.S. Industrial Policy?* Philadelphia: University of Pennsylvania Press.

Williamson, Oliver E. 1975. *Markets and Hierarchies: Analysis and Antitrust Implications*. New York: Free Press.

______. 1979. "Transaction-Cost Economics: The Governance of the Contractual Relations." *Journal of Law and Economics* 22: 233-62.

Wilson, Robert W. 1977. "The Effect of Technological Environment and Product Rivalry on R&D Effort and Licensing of Inventions." *Review of Economics and Statistics* 59: 171-78.

Wolfman, B. 1965. "Federal Tax Policy and the Support of Science." *University of Pennsylvania Law Review* 114: 171-86.

Worley, James S. 1961. "Industrial Research and the New Competition." *Journal of Political Economy* 66: 183-86.

Zaltman, G.; R. Duncan; and J. Holbek. 1973. *Innovations and Organizations.* New York: John Wiley.

About the Authors

BARRY BOZEMAN is professor of public administration and director of the Technology and Information Policy Program at the Maxwell School, Syracuse University. He has published articles on public policy and organization behavior in a variety of journals and is the author of *Public Management and Policy Analysis* (St. Martin's Press, 1981) and coauthor of *New Directions in Public Administration* (Brooks-Cole Publishing, 1983). His papers on science and technology have appeared in such journals as *Research Policy, Public Administration Review, Technological Forecasting and Social Change, Administration and Society, R&D Management, Policy Analysis, Polity, Journal of Public Policy*, and *Policy Studies Journal*. Professor Bozeman is on the editorial boards of *Administration and Society, American Review of Public Administration*, and *The Urban Interest*.

Professor Bozeman has served as a consultant to a variety of state and federal government agencies and as scholar-in-residence at the National Science Foundation's Office of Science Information Service. He has also been on the faculties of the University of Missouri, Virginia Polytechnic Institute and State University (visiting), and Georgia Institute of Technology. He received his Ph.D. in political science from the Ohio State University (1973).

ALBERT N. LINK is professor of economics and department head at the University of North Carolina at Greensboro. He has published widely in the field of economics, especially in the areas of R&D, innovation, and technological change. He is author of *Research and Development Activity in U.S. Manufacturing* (Praeger, 1981) and coauthor of *The Entrepreneur: Mainstream Views and Radical Critiques* (Praeger, 1981). Professor Link's scholarly works have appeared in such journals as the *American Economic Review, The Bell Journal of Economics, Journal of Political Economy, Southern Economic Journal, Decision Science, IEEE Transactions on Engineering Management, Journal of Industrial Economics*, and many more. In addition, he has contributed to several edited volumes, including the forthcoming *Encyclopedia of Economics*.

Professor Link serves on the editorial board of the *Review of Industrial Organization* and has served as a technical consultant to several government agencies and major corporations.

Professor Link holds a B.S. degree (1971) in mathematics from the University of Richmond and a Ph.D. degree (1976) in economics from Tulane University. He was a member of the economics faculty at Auburn University (1976-82) and a visiting member of the public administration faculty at the Maxwell School, Syracuse University, in 1981.